Lecture Notes in Networks and Systems

Volume 1282

The series "Lecture Notes in Networks and Systems" publishes the latest developments in Networks and Systems—quickly, informally and with high quality. Original research reported in proceedings and post-proceedings represents the core of LNNS.

Volumes published in LNNS embrace all aspects and subfields of, as well as new challenges in, Networks and Systems.

The series contains proceedings and edited volumes in systems and networks, spanning the areas of Cyber-Physical Systems, Autonomous Systems, Sensor Networks, Control Systems, Energy Systems, Automotive Systems, Biological Systems, Vehicular Networking and Connected Vehicles, Aerospace Systems, Automation, Manufacturing, Smart Grids, Nonlinear Systems, Power Systems, Robotics, Social Systems, Economic Systems and other. Of particular value to both the contributors and the readership are the short publication timeframe and the world-wide distribution and exposure which enable both a wide and rapid dissemination of research output.

The series covers the theory, applications, and perspectives on the state of the art and future developments relevant to systems and networks, decision making, control, complex processes and related areas, as embedded in the fields of interdisciplinary and applied sciences, engineering, computer science, physics, economics, social, and life sciences, as well as the paradigms and methodologies behind them.

Indexed by SCOPUS, EI Compendex, INSPEC, WTI Frankfurt eG, zbMATH, SCImago.

All books published in the series are submitted for consideration in Web of Science.

For proposals from Asia please contact Aninda Bose (aninda.bose@springer.com).

Alexei L. Semenov · Vadim V. Grinshkun ·
Svetlana N. Dvoryatkina · Vladimir A. Faerman
Editors

Structural and Technological Transformation of Education in the Post-Pandemic Period

Problems and Prospects

Springer

Editors
Alexei L. Semenov
Lomonosov Moscow State University
Moscow, Russia

Svetlana N. Dvoryatkina
Bunin Yelets State University
Yelets, Russia

Vadim V. Grinshkun
Institute of Digital Education
Moscow City University
Moscow, Russia

Vladimir A. Faerman
Control Systems and Radioelectronics
Tomsk State University
Tomsk, Russia

ISSN 2367-3370 ISSN 2367-3389 (electronic)
Lecture Notes in Networks and Systems
ISBN 978-3-031-84038-8 ISBN 978-3-031-84039-5 (eBook)
https://doi.org/10.1007/978-3-031-84039-5

This Springer imprint is published by the registered company Springer Nature Switzerland AG
The registered company address is: Gewerbestrasse 11, 6330 Cham, Switzerland

Preface

In the shadows of the global pandemic, an extraordinary challenge arose, reshaping the very fabric of education as we know it. The COVID-19 crisis compelled educators, students, and institutions alike to navigate uncharted waters, forcing a swift and often tumultuous transition to remote learning. Among the myriad subjects impacted, mathematics—long celebrated for its precision and relevance—faced unique trials and transformations in Russia's global education landscape.

The emergence of *Structural and Technological Transformation of Education in the Post-Pandemic Period* represents a crucial reflection on these experiences. This book offers a comprehensive examination of the multifaceted changes within STEM education, specifically focusing on mathematics and computer science, at various educational levels. The journey through its pages reveals not just the challenges faced but also the resilient spirit of educators and learners who adapted to an ever-evolving environment.

The first part, *Transformation of STEM Education at Pre-University Level*, presents a comprehensive exploration of how educational methodologies are evolving to better integrate technology and innovative teaching strategies in science, technology, engineering, and mathematics (STEM) education for schoolchildren. This section highlights the crucial role of Information and Communication Technology (ICT) in enhancing classroom learning experiences, emphasizing the various stages of digital renewal that schools undergo as they adopt modern educational technologies. It discusses the progressive implementation of digital tools and platforms, illustrating how these advancements facilitate more interactive and engaging learning environments.

A focus is on the development of complex multi-stage tasks designed to assess schoolchildren's understanding and skills in mathematics. These tasks aim to foster critical thinking and problem-solving abilities, providing a more nuanced assessment framework that goes beyond traditional testing methods. The inclusion of hybrid intelligent systems in teaching practices is explored as a means to improve the evaluation of mathematical education quality. These systems leverage artificial intelligence to personalize learning experiences and track students' progress in real time, thereby enhancing educators' capability to support diverse learning needs.

The integration of visualization techniques in blended learning environments is also discussed, particularly in the context of teaching geometry. This approach allows students to grasp abstract concepts more concretely, bridging the gap between theoretical knowledge and practical application. The use of popular science films as educational resources is advocated as a way to incorporate media education technologies into the mathematics curriculum. This strategy not only enriches the learning experience but also connects mathematical concepts with real-world contexts. Moreover, the section examines innovative approaches to teaching geometry through geometric abstractions in real space. This perspective reveals new possibilities for distance learning, making geometry more accessible and engaging for students amidst the challenges of remote education.

Lastly, methodical support for hybrid intellectual systems is addressed, particularly regarding cognitive modeling of engineering-research activities. This aspect emphasizes the importance of recognizing individual thinking profiles to tailor educational strategies that foster deeper understanding and engagement in STEM fields.

The part on the *Transformation of STEM Education at University Level* delves deeply into the challenges and innovations that emerged in mathematics education during the COVID-19 pandemic. It highlights how traditional educational methodologies were altered to accommodate remote learning and digital transformation, which became necessary due to health and safety concerns.

One key theme is the adaptation of distance learning methodologies specifically tailored for mathematics instruction. Educators faced the challenge of delivering complex mathematical concepts in an environment that lacked the immediacy and interaction of in-person classes. This led to innovative strategies for engaging students in remote settings, including the use of virtual classrooms and interactive online tools designed to facilitate collaboration and comprehension. Therefore, the section discusses the imperative for digitalization in mathematics education at technical universities: as the pandemic exposed the limitations of conventional teaching methods, institutions were pushed toward embracing digital resources, revealing both institutional readiness and gaps in technology.

The application of machine learning and data-driven approaches for assessing student performance is another significant focus area. The deployment of intelligent ranking systems to control students' knowledge illustrates how technology can enhance educational assessment, providing more personalized feedback and supporting educators in identifying areas for intervention.

Another practical suggestions investigated in the book is the creative use of mathematics history to enrich the learning experience, particularly in subjects like integral calculus. By weaving historical contexts into lessons, lecturers sought to maintain student interest and enhance understanding, thereby making abstract concepts more relatable and engaging during a time of social isolation.

Finally, the discussion includes the evolution of blended learning models that effectively combine face-to-face interactions with online learning components. The pandemic prompted educators to rethink and redesign their courses, employing information technologies strategically to create a hybrid educational experience that maintains the rigor of traditional STEM education while capitalizing on the benefits of digital tools.

The third section underscores the critical importance of equipping educators with modern pedagogical skills and methodologies necessary for an effective STEM education system. It emphasizes the adaptation of teacher training programs to incorporate blended learning approaches, which integrate traditional face-to-face instruction with digital learning tools. The focus is on preparing future teachers to navigate this digital landscape, fostering their ability to engage students through innovative practices that enhance mathematical understanding and promote computational thinking. Additionally, the section addresses the necessity for transdisciplinary teaching approaches, advocating for an integrated curriculum that merges mathematics and computer science while leveraging diverse technologies to enhance learning experiences.

Furthermore, the section discusses the evolution of teaching economic disciplines within this digital framework, highlighting the effective use of digital technologies in delivering content during distance learning scenarios. It recognizes the role of technology in not only facilitating instruction but also in assessing student competencies, especially exemplified during the unique challenges posed by the pandemic. By emphasizing these contemporary methodologies and technologies, the section aims to empower future educators, ensuring that they are well prepared to foster a dynamic and engaging learning environment that meets the demands of a rapidly changing educational landscape.

In concluding this exploration of the structural and technological transformations that have shaped education in the post-pandemic period, we reflect on the profound shifts that have redefined teaching and learning landscapes worldwide. The challenges posed by recent global events have not only tested our resilience but have also catalyzed innovative approaches and solutions within the educational framework. While we celebrate the achievements of educators, institutions, and students alike, we recognize the critical role that continuous adaptation and inquiry play in nurturing the minds of future generations.

We extend our sincere gratitude to the editorial and publishing team at Springer Nature, whose dedication and professionalism have been instrumental in bringing this collective work to fruition. The insights provided by our academic reviewers have enriched the discourse and enhanced the rigor of our contributions, embodying the collaborative spirit of the international academic community, particularly in the field of mathematics education.

To our readers, especially those of you who are currently students with aspirations to become teachers in STEM disciplines, we encourage you to embrace the dynamic nature of education. The journey ahead is filled with opportunities to inspire, innovate, and influence the next wave of learners. Your commitment to growth and learning will shape the future of education and empower countless individuals to navigate an increasingly complex world.

Moscow, Russia — Alexei L. Semenov
Yelets, Russia — Svetlana N. Dvoryatkina
Yaroslavl, Russia — Eugeny Smirnov
Moscow, Russia — Vadim V. Grinshkun
Tomsk, Russia — Vladimir A. Faerman

The original version of the book has been revised. A correction to this book can be found at https://doi.org/10.1007/978-3-031-84039-5_18

Contents

Transformation of STEM Education at University Level

Teachers Training for STEM Education System

Transformation of STEM Education at Pre-University Level

Innovative ICT-Supported Teaching and the School's Digital Renewal Stages

Irina Dvoretskaya and Alexander Uvarov

Abstract Evolving digital technologies are infiltrating schools nowadays. This process is taking place by stages that are viewed as the schools' digital renewal process (SDRP), the complex and multidimensional one. It includes changes in the educational environment (physical and virtual), the educational process, and the way how school operates. The use of innovative ICT-supported teaching (IITST) can be considered as one of the key indicators of a school' digital renewal process for measurement of its progress. The aim of the study is to evaluate the status of innovative IITST within the framework of SDRP. During the monitoring of the digital transformation of Russian general education in 2021 data on IITST ware collected. More than 500 schools from all regions of Russia participated in the study. The data collected show that schools in different regions of the country are unevenly progressing along the path of digital renewal. The report presents results obtained during the monitoring and discusses the reasons for the uneven movement of schools along the path of digital renewal in Russia.

Keywords ICT in education · Digital transformation of education · School' digital renewal process · Monitoring of digital transformation · Innovative ICT-supported teaching · Digital capacity of schools

I. Dvoretskaya (✉) · A. Uvarov
Higher School of Economics, 20, Myasnitskaya St, 101000 Moscow, Russia
e-mail: idvoretskaya@hse.ru

A. Uvarov
e-mail: auvarov@hse.ru

A. Uvarov
Institute of Cybernetics and Educational Computing, Federal Research Centre Computer Science and Control, Russian Academy of Sciences, 40, Vavilova St, 119333 Moscow, Russia

A. L. Semenov et al. (eds.), *Structural and Technological Transformation of Education in the Post-Pandemic Period*, Lecture Notes in Networks and Systems 1282,
https://doi.org/10.1007/978-3-031-84039-5_1

1 Introduction

During the last decades, schools were experienced three waves of ICT spread: computerization, early and late informatization of education. The fourth wave of changes is approaching—the digital transformation of education. These waves represent stages in the school's digital renewal process (SDRP). While there is a lack of approaches and tools to manage ICT integration in education qualitatively [1], decision-making relies mostly on statistical quantitative data of digital infra-structure changes, which often ignore qualitative changes [2, 3]. To address it, schools' survey data that describes changes in processes and their meaningful interpretation can be used for evidence-based educational policy development. In this relate qualitative interpretation of changes based on survey data analysis would help to shape policy documents and to prepare practical recommendations for schools and local educational authorities.

We propose using of innovative ICT-supported teaching (IITST) as a key indicator of a SDPR progress. Our study examines this in the Russian context and align the stages of innovative work against SDRP stages to reveal the current status.

2 Literature Review

2.1 *SDPR as a Framework to Measure the ICT Use in Education*

SDRP is a dynamic process that tackle all aspects of education: teaching and learning, educational environment, teacher preparation, and professional development. In order to take into account the different paces schools move along the path of digital renewal, the periodization of SDPR should be done.

It is possible to describe the process of SDRP as four main stages:

Computerization

Schools at this stage focus on equipping with computers and other digital devices. ICT is used mainly for computer literacy development in the computer labs as well as for routine tasks in school management and traditional teaching. As teachers start use ICT to update teaching and learning, conditions will evolve to advance to the next stage of digital renewal.

Early informatization

Schools at this stage focus on the use of ICT in the study of separate subjects. ICT usafe is expanding beyond computer classes. Digital learning materials and instruments are emerging for teaching.

Late informatization of education

It is the third stage of SDPR, school's educational environment is saturated with digital technologies. Implementation and development of innovative ICT-supported teaching and learning is underway. Today, there are a significant number of schools from economically developed countries at this stage of SDPR maturity.

Digital transformation of education

Only some schools are entering the digital transformation stage now [2]. Within digital transformation implementation of personalized mastery-based learning is supported by cloud technologies, Personal Learning Platforms (PLP), digital learning resources collections, and intelligent tutoring systems (ITS). The trends are being carried out today in many countries globally.

For practical reasons, it can be considered that the transition of a group of schools to the next stage has begun if at least 10% of schools ("innovators" and a significant part of "pioneers", in Rogers' terminology [4]) have reached the state that is the initial for a new stage. The transition to a new stage can be considered as accomplished after the main parts of school in the group/territory ("first" and "second majority", according to Rogers [4]) have reached a new state.

This division along stages can be used as a basis for planning of the school development as well as determining the mechanisms and volumes of state support schools need. SDRP maturity level can inform strategical goals and choose next steps using available recourses. That is why the development of indicators that can be used for maturity level of SDRP assessment becomes more and more actual.

2.2 *IITST as a Key Process of SDRP*

The digital transformation of education has been accelerating, marked by an increase in both the number and complexity of indicators [5]. This complexity demands sophisticated interpretation methods. There is a growing need for meaningful analysis of the survey results [5]. However, there is a noticeable deficit guiding the study of digital renewal and digital transformation of schools [1]. The use of IITST can be considered as one of the key indicators of a school' digital renewal process for measurement of its progress.

Research show that innovative ways of teaching drive the modernization of education system [6]. Usually some types of work are contemplated: blended learning; teaching in small groups; project-based learning, individual and group work using digital educational resources; creation of digital artefacts; research projects done by students (individually and in groups), personalized learning [6, 7]. ICT-supported practices can vary significantly from school to school [8] and their taxonomy is large and poorly defined. The presence of any innovative ICT-supported educational activities usually declared by teachers regardless of transformation of teaching and learning. What shapes innovative ICT-supported teaching more is a supportive school environment and schools' change management practices [9].

3 Materials and Methods

To categorize schools according to the use of IITST, we used the combined procedure. The procedure integrates quantitative survey data with qualitative maturity level descriptions. We adapted the combined technique described previously [10, 11] to analyze data from the Monitoring of digital transformation of general education (MDTGE), collected at the end of the 2020/2021 school year. Our sample included 509 schools with responses from at least five teachers per school, involving 2,278 principals and 14,938 teachers. The technique was modified to a 4-point Likert scale with a "not sure" option, using two specific statements from the MDTGE surveys (Table 1).

Looking ahead of a result of clustering on our data, schools were distributed into 15 clusters described below based on computed characteristics (Table 2).

Continuation of Table 2.

Table 1 Diad of statements from MDTGE

School teachers' statement	School leaders' statement
I am using digital technologies to try new ways of teaching	We support teachers in testing new ways of delivering digital education

Table 2 Cluster descriptions

No	Number of schools	Description
1	2	There are certain groups of teachers that use the use of innovative ICT-supported teaching and learning but only some school leaders support them
2	1	School leaders support innovative ICT teaching not consistently, and teachers consistently use these practices
3	4	Individual school leaders and teachers have started using innovative ICT-supported teaching, but it goes neither systematically nor consistently
4	3	Individual school leaders and groups ot teachers have just begun to adopt innovative ICT-supported teaching and can plan and organize it
5	3	School leaders organize the use of innovative ICT-supported teaching and learning through the development of plans and supportive measures
6	6	School leaders plan and organize the use of innovative ICT-supported teaching and learning. Most teachers of the school use traditional ways of teaching
7	9	School community is not aware that they need ICT-supported innovative teaching and learning
8	7	Individual school leaders begin to take steps towards the use of innovative ICT-supported teaching and learning, but not systematic

No	Number of schools	Description
9	77	School leaders plan and organize ICT-supported teaching and learning. Some teachers also use innovative ICT-supported teaching
10	4	School leaders plan and organize the use of innovative ICT-supported teaching. Some teachers also use innovative ICT-supported teaching
11	115	Some of school leaders are aware that there is need to support innovative teaching, so they plan and organize ICT-supported teaching and learning. Only individual teachers use innovative ICT-supported teaching
12	1	Some of school leaders plan and organize ICT-supported teaching and learning, while teachers consistently use innovative practices
13	52	Both leaders and teachers lack awareness of the need to support innovative ICT-supported teaching. However, some teachers proactively use ICT for teaching
14	218	Some of school leaders recognize the need to support innovative teaching. Individual teachers use innovative ICT-supported teaching
15	7	Some of school leaders recognize the need in innovative teaching. There are distinct groups of teachers who are also taking steps in this direction

4 Results and Discussion

Matching clustering results against RISC maturity levels, we observed the distribution of schools. In 11.9% of schools' teachers prefer mostly traditional ways of teaching and learning, while school management not prioritizing innovative ICT-supported teaching. In contrast, 88.1% of schools have recognized the need to use innovative ICT-supported teaching and learning, they organize their educational practices accordingly (Table 3).

In all schools innovative ICT-supported teaching and learning is at least introduced.

School at the "Traditional" level of the RISC scale means substitute paper tool by digital ones without significant changes in teaching [12]. In most schools (the level of

Table 3 School cluster structure against RISC maturity levels

Level of RISC scale	Clusters	Number of schools	% of schools
Traditional	7, 13	61	11.9
Awareness	1, 2, 3, 6, 8, 11, 12	360	70.5
Understanding	14, 15	81	15.9
First implementation	4, 9, 12, 5, 10	7	1.4
Usage	–	–	–
Elaboration	–	–	–

"Awareness"), ICT-supported teaching is ad-hoc and non-systematic, supporting by individual teachers and/or school leaders. The large number of schools at this level indicates early-stage digital renewal, though the distribution is heterogeneous. The largest clusters (11 and 14) are led by proactive school management, while clusters 1 and 15 feature active groups of teachers experimenting with ICT.

Schools at the "Understanding" and "First implementation" levels show manageable innovative ICT-supported teaching efforts by school teams, with supportive administration knowledgeable about digital renewal. Only 1.4% of schools systematically use innovative ICT-supported teaching, which aligns with their late informatization stage. The goal of teaching and learning renewal is relevant for a few schools, and some "First implementation" schools might be more advanced, possibly at the "Usage" or "Elaboration" levels, which requires careful examination.

5 Conclusion

Empirical validation of SDPR key indicators enriches its theoretical framework and submits it to practical recommendation preparation to schools.

In this study, we assess the situation with innovative ICT-supported teaching in a relatively large sample of schools with diverse geographical characteristics. Our findings help identify systemic factors of innovative ICT-supported teaching. Comparing our data with the results of the previous study using SELFIE data from 2017/18, we reveal minimal systemic changes over three years as only 1.4% of schools are systematically using innovative ICT-supported teaching, corresponding to the "Late informatization" level. There is a disparity among schools within the same maturity level of the SDRP. A more accurate picture would emerge from monitoring data with essential (key) indicators of digital renewal.

This study has several limitations. Firstly, our study used a non-representative sample with a target population of Russian schools. Given the availability of an online survey platform in MDTGE, it is reasonable to consider that the national landscape of innovative ICT-supported teaching is more challenging. This would be a valuable time to check this hypothesis during the next waves of MDTGE in Russia.

Acknowledgements The reported study was funded by RFBR, project number No 19-29-14167.

References

1. Dvoreckaya IV, Uvarov AY, Vihrev VV (2020) Models for Renovating General Education in an Emerging Digital Environment. TORUS-Press
2. Lee M, Broadie R (2016) A taxonomy of school evolutionary changes. Igarss 2014. Broulee Australia

3. Uvarov AY (2020) Digital transformation and scenarios for the development of general education. Higher School of Economics, Moscow 16
4. Rogers EM, Singhal A, Quinlan MM (2019) Diffusion of innovations. An Integrated Approach to Communication Theory and Research, 3rd edn. Routledge, Abingdon
5. Ređep NB, Balaban I, Žugec B, Čalopa MK, Divjak B (2019) Framework for digitally mature schools. Eur J of Open, Dist and E-Learn (EURODL) 22(1):360–371
6. Law N, Pelgrum WJ, Plomp T (2008) Pedagogy and ICT Use. In: CERC studies in comparative education, vol. 23. IEA, Hong Kong
7. Punie Y, Bocconi S, Kampylis P (2012) In-depth Key Elements for Developing Creative Classrooms. In: eLearning papers, vol 30, pp 1–13. In-depth
8. Brečko BN, Kampylis P, Punie Y (2014) Mainstreaming ICT-enabled innovation in education and training in Europe: Policy actions for sustainability, scalability and impact at system level. Joint Research Centre, Luxembourg
9. Shear L, Gallagher L, Patel D, Fullan M (2011) Innovative Teaching and Learning research: 2011 Findings and Implications. SRI International, Menlo Park
10. Dvoreckaya I, Uvarov A (2020) Assessing the school's use of innovative, digitally supported learning work based on quantitative data. Domestic and Foreign Pedag 2(66):29–45
11. Begicevic RN, Balaban I, Zugec B (2021) Assessing digital maturity of schools: framework and instrument. Technol, Pedag and Ed 30(5):643–658
12. Kampylis P, Devine J, Punie Y, Newman T (2016) Supporting schools to go digital: From a conceptual model towards the design of a self-assessment tool for digital-age learning. Paper presented at the ICERI2016 Conference. Seville, Spain, 14–16 November 2016

Complex Multi-Stage Tasks for Testing Schoolchildren in the Mathematics Course

Eugeny Smirnov

Abstract The modern education system strives for an individualised approach, allowing each student to unlock his or her potential. Hybrid intelligent systems play a key role in this process. These systems offer level adaptation. Complex knowledge is presented according to the individual needs of each student. They aim at developing creativity. Students are given the opportunity to realise their potential in solving a variety of problems. Hybrid systems encourage multiple goals. Students become participants in a multifaceted learning process and have the opportunity to choose the direction of their activity. There is a synthesis of two disciplines, which allows solving complex problems with the help of computer modelling. As a result, students self-organise. They become active participants of the educational process, form their own way of development. They master not only knowledge, but also methods of scientific cognition, which allows them to apply modern achievements in science. The aim of this work is to create a system that will allow students to develop their scientific potential in the conditions of digital transformation. The basis of this work is a hybrid intellectual environment. Such an environment provides the development of hierarchical databases and fundamental problem sets for students' research activities. Intellectual management within such an environment provides directed development of project-research activity.

Keywords Teaching mathematics · Intelligent systems · Mathematical modeling · Computer simulation

1 Introduction

Mathematical education in the modern world is undergoing on significant changes. The effectiveness of modern society functioning and high-tech trends implementation in production and technology require the education of creative individuals in the

E. Smirnov (✉)
Yaroslavl State Pedagogical University, 108, Respublikanskaya St, 150000 Yaroslavl, Russia
e-mail: smiei@mail.ru

A. L. Semenov et al. (eds.), *Structural and Technological Transformation of Education in the Post-Pandemic Period*, Lecture Notes in Networks and Systems 1282,
https://doi.org/10.1007/978-3-031-84039-5_2

conditions of self-development of their scientific potential. The student's personality has changed in the modern period in the direction of taking into account and realizing her interests, preferences and social activity. At the same time, the increasing intensity of sciences development, modern achievements in applied mathematics (fractal geometry, fuzzy sets and fuzzy logic, neural networks and artificial intelligence, theory of information encoding and encryption, cellular automata, Schwartz cylinder and cone) dictate the need to integrate the science and education as a fundamental paradigm for the development of school mathematical education. These trends allow us to presence of rich information and educational environment, the deployment of hierarchical bases and complexes of practice-oriented tasks, the availability of effective feedback and monitoring of scientific potential growth. The *leading idea* for defining the research objectives is as follows: the manifestation key aspect of synergetic effects phenomenon and self-organization in mathematics development can be the digital transformation of each student's scientific potential. These trends lead to an actualization of complex level knowledge in rich information and educational environment of intellectual management of student's research activities (including activity on the basis of generalized constructs adaptation of modern achievements in science) in the course of "problem areas" of mathematics mastering.

Thus, the *main objective* is to develop a technology for digital transformation of complex knowledge content with hybrid intelligent systems support. It is realized during the implementation of research activity and growth of each student's scientific potential stages by means of mathematical and computer modeling symbiosis.

2 Survey Methodology

The implementation of this technology is aimed at developing complex knowledge in students. This is achieved through the application of mathematical and computer modelling in an information-rich educational environment. Intelligent systems support this process, contributing to the growth of scientific potential of students [1]. The modern world, characterised by complexity, diversity and unpredictability, requires learners to be able to adapt to constantly changing conditions. This is associated with the development of the ability to self-organisation, effective thinking and formation of personal qualities (S. Bier [2], N. Wiener, J. von Neumann). Researchers (S.P. Kurdyumov, G. Haken [3], K. Mainzer [4], A.N. Podyakov [1], V.S. Stepin [5]) emphasise the importance of complex knowledge assimilation, overcoming difficulties in the learning process and forming a unified picture of the world. Motivation and independence of learners in the context of interaction and interrelation are of special importance. The process of cognition, in turn, becomes a key element in understanding complexity and development [6].

2.1 Related Works on Complexity of Knowledge

Complex mathematical knowledge can be applied or practice-oriented. Its study is based on the combination of mathematical and computer modelling. Examples of such knowledge include: fractal geometry [6, 7], graph theory, fuzzy sets and logic, information encoding and encryption, stochastic methods for solving optimisation problems. Intelligent systems can contribute to the formation of complex knowledge, improving the quality of project and research activities of students. This is possible with the active participation of the teacher and relevant professional competences [9].

2.2 Intelligent Systems Typology

Artificial neural networks (ANN) have been actively used in education in recent years. Their application makes it possible to solve the problems of personalisation and automation of learning. Research related to ANNs in education focuses on adaptive testing, personalised learning, intelligent control of the learning process and intelligent learning support [6, 10–16]. Intelligent control in mathematics education of students is based on the use of ANNs in the openness and synthesis of mathematical and computer modelling [17]. This study aims to develop technology for intelligent control of students' mathematical activities. It is based on holistic structures with hierarchical content built on modern topics of mathematics and facilitating the deployment of individual educational routes in an enriched information and educational environment.

2.3 Data and Methods

We propose the following *typology of design levels and student's research activity in hybrid intellectual environment* [9]:

- *Search-reproductive stage (PR) is* a period when the learner is interested in the material being studied and shows self-actualisation ('I am interested in this'). At this stage there is an active inclusion of the personality in the learning process, the aspiration for self-determination and self-organisation develops. The learner begins to master the principles and styles of scientific thinking. He learns about such methods of cognition as: induction (generalisation from particular to general), deduction (deduction of particular from general), insight (sudden insight), analogy (comparison), inversion (reversal of order), and ancipation (anticipation). This enables him/her to understand the material he/she is learning in depth and to apply it to real-life situations.

- *Empirical stage (E)* is characterised by active involvement of the learner in practical activities. At this stage, the learner has a desire to apply the acquired knowledge and test it in practice ('what can I do?'). The learner performs empirical tests, constructs visual models using methods of experimental mathematics, mathematical and computer modelling. He/she explores particular manifestations of the essence of complex knowledge, relying on the actualisation of cognitive independence, development of competences and formation of personal qualities. This allows him to deeply understand and consolidate theoretical knowledge, acquire practical skills and develop creative thinking.
- *Theoretical stage (T)* is characterised by deep immersion of the learner in the process of cognition and development of the ability to independently manage learning ('I know how to manage the process'). At this stage, the learner gains experience in designing and organising technologies for mastering research procedures. The learner learns about the latest achievements in science and learns to adapt them to practical tasks. He/she learns the essence of generalised constructs of complex knowledge and applies them to solve specific problems. As a result, the student develops deep theoretical knowledge and practical skills that enable him/her to confidently navigate the complex world of modern science.
- *Creative stage (TV)* is developed by results evaluation of empirical verification and characterized by self-development of personality ("I can do something new").

The following Fig. 1 shows the content and structure of hybrid neural environment to support the growth of student's scientific potential [18].

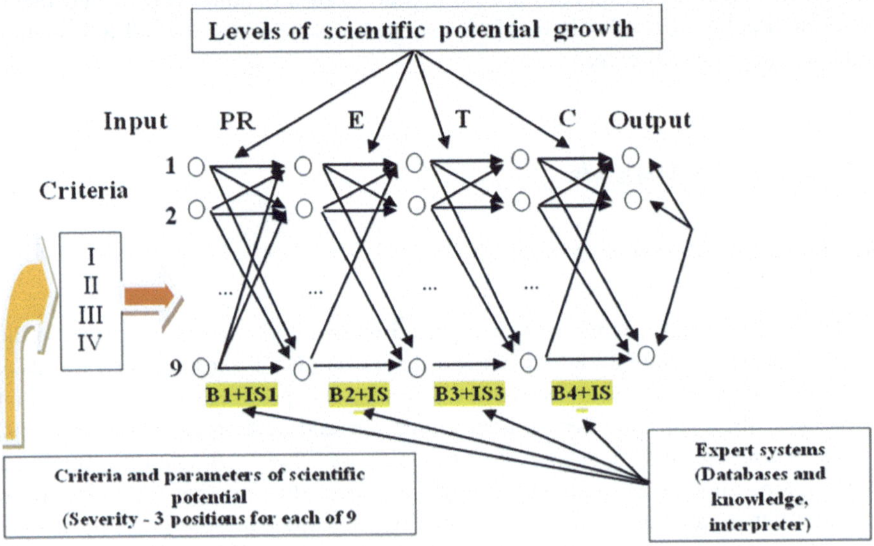

Fig. 1 Dynamics of parameters variation and success levels and research activity quality in the layers of hybrid neural network

Human interaction with the world and other people plays a key role in human development. It activates a person's inner capacities, contributing to self-knowledge, self-regulation and self-actualisation. These processes are the foundation for personal self-development, enabling a person to fulfil his or her potential and achieve his or her goals. The success of research activity largely depends on the individual educational trajectory. This trajectory is based on the learner's personal preferences and achievements, in the context of an information-rich environment. This individual approach allows synthesising personal self-organisation and mathematics education, which facilitates the assimilation of complex knowledge, including modern achievements in science. This makes the educational process more effective and allows each learner to develop in accordance with their individual characteristics and goals.

3 Results and Discussion

We will highlight a number of *technological stages of founding procedures deployment in adaptation processes of* knowledge complexity (modern achievements in science) to school mathematics:

1. *Mastering the standards and samples* of complex knowledge generalized constructs and diagnostic procedures results of specific manifestations of generalized construct essence [19–21];
2. *Actualization of motivational field* in the development of complex knowledge generalized construct [22];
3. *Selection of "problem areas" and research tasks* to actualize the deployment of individual educational trajectories for small groups [23].
4. *Multiple goal-setting of the content and processes* in the studying of complex knowledge generalized construct in "problem zone" of mastering mathematics, taking into account of student's interests and readiness [24].
5. *Readiness for discussions and multiplicity of solutions to the problem*; identification of criteria for selection, formulation and search for solutions to research practice-oriented tasks based on diagnostic information systematized in the form of founding hierarchical complexes of subtasks in complex knowledge generalized constructs [25];
6. *Construction a creative environment* in mastering process of generalized construct essence (motives realization in cognitive activity, stimulation of success situation and availability awareness of pedagogical and informational support (including hybrid intellectual systems) [26, 27];
7. *Effectiveness of mathematical, informational, natural science and humanitarian cultures dialogue* based on computer and mathematical modeling of components and stages of generalized construct adaptation in "zones of modern achievements in science" to school mathematics [8, 28];

8. *Actualization of synergy attributes* (bifurcations, attractors, fluctuations, pools of attraction) in research process, visual modeling and founding of complex knowledge generalized construct [22, 29].

3.1 *Cluster of Student's Design Content and Research Activities*

Example 1. Generalized construct of complex knowledge: Fuzzy sets and fuzzy-logic.

The problem of this cluster is the difficulty of qualitative mastering by students of applied approaches to solving problems of economics, engineering and communication by traditional means of computer modelling. The generalised construct of the underlying cluster is realised through a didactic model (the model includes fuzzy sets and fuzzy logic). The didactic model used in the student's project-research activity consists of 4 phases (Fig. 2):

- An initial level of essence mastering (search and reproductive level (**PR**) is fuzzy sets at the intuitive and visual levels and historical and genetic levels;

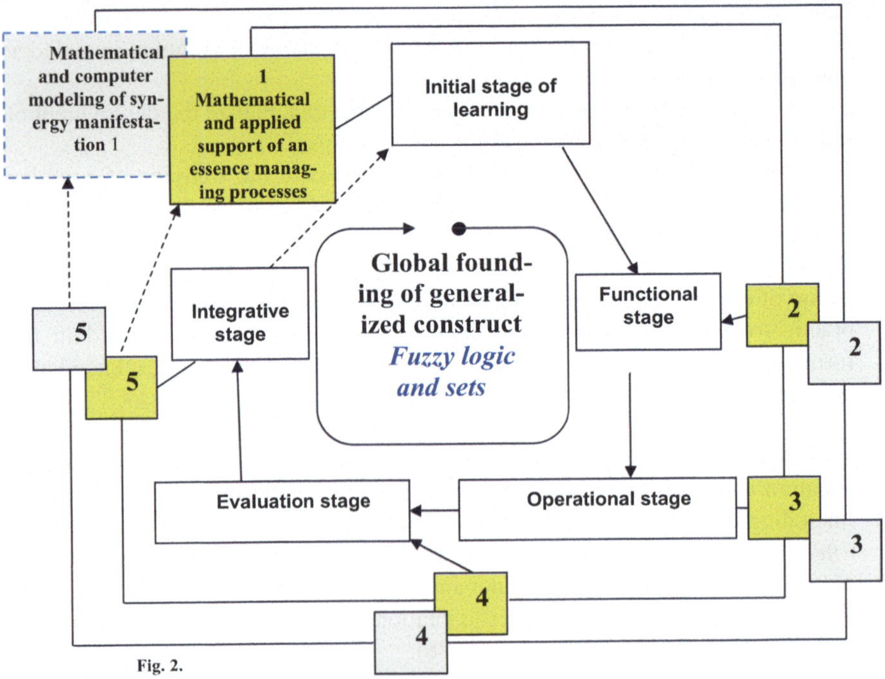

Fig. 2 Founding cluster of generalized construct essence (fuzzy logic and fuzzy sets)

- Functional and operational stages (empirical level—**E**);
- Stage of awareness and generalization (**T**));
- Integrative stage (creative level (**C**)).

Each stage is integrated with three founding spirals by equipping means of deployment and adaptation processes of generalized construct essence.

At the end of first level implementation by presentation of research directions and the essence manifestation of complex knowledge generalized construct, each student is offered opportunities for variable design and research activities within the intellectual environment. Namely, 3–4 areas of research related to selected generalized construct are being updated (in our example, fuzzy sets and fuzzy logic):

- Modeling of ecological and economic processes and phenomena based on fuzzy sets;
- Fuzzy inference algorithms in simulation modeling;
- Mathematical foundations and algorithms for fuzzy modeling of complex systems.

At the end of first level implementation each student has a variable choice of one of the 5 directions: sign-symbolic, figurative-geometric, verbal, concrete-activity and historical-genetic. This technology should preferably be preserved until the results of design and research activities and student's scientific potential growth.

3.2 Discussion

The modern world is characterised by the rapid development of technology, which increasingly influences various spheres of life, including education. Educational institutions are increasingly using modern technological resources, which provide ample opportunities for communication and use of complex educational programmes. This paper proposes an architecture for an intelligent tutoring system that helps to manage the design and research activities of students. The system takes into account the individual characteristics of each student: their personal preferences, intellectual abilities and experience in mathematics. The proposed system has a number of advantages over similar systems developed earlier. It contributes to the growth of students' scientific potential, develops their creativity and critical thinking. The system helps students to master complex knowledge including areas such as fuzzy sets and logic, fractal geometry, cellular automata, information coding and encryption, chaos and catastrophe theory, and non-Euclidean geometry. The system provides students with research tasks, instructions and rules for organising design and research activities, taking into account their level of scientific potential. This creates conditions for effective management of students' research activities and promotes their personal growth and professional development.

4 Conclusion

Thus, methodological, substantive and technological foundations for the design and actualization of digital transformation processes of hierarchical databases deploying and founding complexes development of multi-stage mathematical and information tasks are identified as the results of goal setting. The content, computer design and technology of generalized constructs learning for identifying and adaptation its essence at "problem areas" of school mathematics by means of intelligent systems are developed. In this direction, the generalized stages and learning content of fuzzy sets and fuzzy logic construct in extracurricular activities with the effect of scientific potential growth of each student are considered. The possibilities and necessity of identifying and characterizing of bifurcation points, attraction basins, computational procedures and fluctuations of state parameters, computer design and side results of generalized constructs learning are determined. Similar studies can be carried out to adapt the generalized constructs of complex knowledge: nonlinear dynamics of an areas growth for polyhedral complexes when grinding triangulations of lateral surface "area" of Schwartz cylinder and cone, elements of fractal geometry. This direction reflects the content of new didactics using intelligent systems of teaching mathematics.

Acknowledgements The research was founded by RFFI Project No 19-29-14009.

References

1. Podyakov AN (2006) Methodological foundations for the study and development of research activities. Res Activ of Stud in the Modern Ed Space 1:61–75
2. Beer S (1972) Brain of the Firm: The Managerial Cybernetics of Organization. Allen Lane the Penguin Press, London
3. Haken H (1996) Principles of brain functioning. A Synergetic Approach to Brain Activity. Behavior and Cognition. Springer, Berlin
4. Mainzer K (2009) Complex system thinking. Matter, mind, humanity. New synthesis. Librokom, Moscow
5. Stepin VS (2011) Historical types of scientific rationality in their relation to the problem of complexity. Synergetic paradigm. Synergetics of innovative complexity. Progress-Tradition, Moscow, pp 37–46
6. Morin E (2007) Education in the future: seven urgent tasks. Synergetic paradigm. Synergetics of education. Progress-Tradition, Moscow, pp 24–96
7. Mandelbrot BB (2002) Fractal geometry of nature. Institute of Computer Research, Moscow
8. Zadeh LA (2012) Computing with Words: Principal Concepts and Ideas (Studies in Fuzziness and Soft Computing). Springer, Berlin
9. Dvoryatkina SN, Smirnov EI, Tikhomirov SA (2020) Self-organization technology of student's mathematical activities based on intelligent management. Perspect of Sci and Ed 45(3):77–86
10. Adebayo AO, Oyewale AE (2015) Intelligent learning management system starters. Int J of Engineer Res and Develop 11(8):53–64
11. Averkin A, Yarushev S (2020) Neural networks in semantic analysis. Open Semantic Technol for Intellig Syst 4:133–136

12. Binh HT, Duy BT (2017) Predicting students' performance based on learning style by using artificial neural networks. Paper presented at the 9th International Conference on Knowledge and Systems Engineering (KSE), Hue, Vietnam, 19–21 Oct 2017, pp 48–53
13. Johanes P, Lagerstrom L (2017) Adaptive Learning: The Premise, Promise, and Pitfalls. Paper presented at the 124th ASEE Annual Conference and Exposition. Ohio, Columbus, p 27538
14. Hadzhikolev E, Yotov K, Trankov M, Hadzhikoleva S (2019) Use of neural networks in assessing knowledge and skills of university students. Paper presented at the 12th annual International Conference of Education, Research and Innovation, pp 7474–7484
15. Kashapov MM (2021) Explication functions in terms of professionalization of thinking. Bull Yarosl State Univ Named After PG Demidov 15(1):100–109
16. Prigogine IR, Nikolis G (2017) Cognition of the Complex. Lenand, Moscow
17. Dvoryatkina S, Khizhnyak A, Smirnov E, Shcherbatykh S, Zhuk LV (2021) Open Innovation model of student's research activities (Running title: Open Innovation Model). J of Teach Ed for Sustainability 23(2):82–95
18. Smirnov E, Dvoryatkina S, Shcherbatykh S (2020) Parameters and structure of neural network databases for assessment of learning outcomes. Int J of Criminol and Sociol 9(83):1638–1648
19. Ainley J, Margolinas C (2015) Accounting for student perspectives in task design. In: Watson A, Ohtani M (eds) Task design in mathematics education. Springer International Publishing, Switzerland, pp 115–142
20. Bikner-Ahsbahs A, Janßen T (2013) Emergent tasks–spontaneous design supporting in-depth learning. In: Watson A, Ohtani M, Ainley J, Bolite Frant J, Doorman M, Kieran C, Leung A, Margolinas C, Sullivan P, Thompson D, Yang Y (eds) Proceedings of ICMI Study 22: Task Design in Mathematics Education. International Commission on Mathematics Instruction, Oxford, pp 153–162
21. Boaler J (1993) Encouraging the transfer of 'school' mathematics to the 'real world' through the integration of process and content, context and culture. Ed Stud in Math ics 25(4):341–373
22. Chaira T (2019) Fuzzy Set and Its Extension: The Intuitionistic Fuzzy Set. Wiley & Sons Ltd
23. Lithner J (2017) Principles for designing mathematical tasks that enhance imitative and creative reasoning. ZDM Mathematics Education
24. Dvoryatkina S, Smirnov E, Lopukhin A (2017) New opportunities of computer assessment of knowledge based on fractal modeling. Paper presented at the 3th International Conference on Higher Education Advances, HEAd/17. Universitat Politecnica de Valencia, Valensia, pp 854–864
25. Karp RM (1993) Combinatorics, complexity, randomness: lectures of the Turing Prize laureate. Mir, Moscow, pp 498–521
26. Kolesnikov AV (2017) Artificial heterogeneous systems supporting decision-making. Artif Intellig Probl 4(1):16–27
27. Martyushev N, Shutaleva A, Malushko E, Nikonova Z, Savchenko I (2021) Online communication tools in teaching foreign languages for education sustainability. Sustainability (Switzerland) 13(19):11127
28. Smirnov E, Dvoryatkina S, Martyushev N, Shcherbatykh S (2023) Software package to support students' research activities in the hybrid intellectual environment of mathematics teaching. Mathematics 11(4):952
29. Shutaleva A, Starostin A, Nikonova Z, Savchenko I (2022) Migration potential of students and development of human capital. Educ Sci 12(5):324

Hybrid Intelligent System of Teaching Schoolchildren as the Dominant Tool for Assessing the Quality of Mathematical Education

Svetlana Dvoryatkina, Larisa Zhuk, and Sergey Shcherbatykh

Abstract The COVID-19 pandemic has had a negative impact on all spheres of life, including economic, political, social and professional. However, the field of education can confidently be called one of those on which the pandemic has had the greatest impact on a global scale. The education sector has become one of the few prepared to transfer most of its processes to the online mode. The experimental schools were able to function effectively and even strengthen themselves in the conditions of the global crisis, in which the educational process is built using hybrid intelligent learning systems. The article reveals the features of the functioning of such systems, the formation of knowledge bases and tools of expert systems, methods, algorithms, methods that provide the processes of operational training, control and evaluation of knowledge, competencies and procedures in the information and educational intellectual environment. Significant results include the conclusions of a pilot experimental study to determine the effectiveness and efficiency of the functioning of hybrid intelligent learning systems in the conditions of distance and blended learning. Expert evaluation of the results of the state unified examination in mathematics confirmed the high quality of subject-based learning in experimental classes based on a hybrid intellectual system developed and implemented in the educational process to support the educational and research activities of schoolchildren.

Keywords Teaching mathematics · Intelligent systems · Unified state exams · Expert assessment in education · Retained knowledge

1 Introduction

The crisis situation caused by the COVID-19 pandemic has led to large-scale changes in many social systems, including in the rapidly developing field of education. The forced transition to mass distance learning in the context of the spread of coronavirus

S. Dvoryatkina (✉) · L. Zhuk · S. Shcherbatykh
Bunin Yelets State University, Kommunarov 21.1, Yelets, Russian Federation
e-mail: prorektor-nr@elsu.ru

A. L. Semenov et al. (eds.), *Structural and Technological Transformation of Education in the Post-Pandemic Period*, Lecture Notes in Networks and Systems 1282,
https://doi.org/10.1007/978-3-031-84039-5_3

infection has become a serious challenge for educational systems around the world, required significant structural and substantive transformations of didactic models based on new communication mechanisms of students and teachers, new forms of evaluation procedures and analysis of their results.

Currently, the Russian Federation has a unified system for assessing the quality of education, which allows monitoring the knowledge of students at different levels of education, promptly identifying and solving problems of the education system in the context of subjects, schools and regions. This system is multi-level, consisting of several evaluation procedures:

- In-school assessment of the quality of education, focused on the diagnosis of the effectiveness of the educational organization (educational achievements of students, the quality of the development and implementation of educational programs, the quality of the support of the educational process, the effectiveness of management);
- State exams presented by the main state exam in the 9th grade and the unified state exam in the 11th grade, which is the main form of the state final certification and at the same time the form of entrance tests for admission to universities of the Russian Federation;
- National studies of the quality of education conducted since 2014 in various academic subjects in order to monitor the current state of the education system, timely adjustments and the formation of programs for its long-term development;
- International monitoring studies (PISA, TIMSS, PIRLS, TALIS, etc.).

The experience of mass distance learning during the COVID-19 pandemic has shown the imperfection of the traditional system for assessing the quality of education. First of all, this affected the intra-school assessment procedure, which, when face-to-face interaction was impossible, made it difficult to receive operational feedback, and acted as a serious barrier in assessing the educational achievements of students. The purpose of this study is to develop a new diagnostic tool for assessing the quality of mathematical training of schoolchildren, which provides an opportunity for continuous diagnostics of the educational process in the conditions of e-learning: tracking the student's progress in achieving set results, determining the current state of learning and possible long-term development. As such a tool, we consider an adaptive learning system based on artificial intelligence methods.

2 Potential of Adaptive Systems During a Pandemic

The COVID-19 pandemic has increased interest in the use of artificial neural networks in education. The introduction of platforms and online services with intelligent assessment of knowledge was actively taking place in foreign educational institutions. Popular platforms include MATHia by Carnegie Learning, Yixue Education, Wayang Outpost, Century Tech, Math-u-See, Education Perfect and others. In Russia during the pandemic period, electronic resources were also actively used by both students and teachers. Among the most actively used resources, the following can be distinguished:

- Information resources and platforms providing access services to educational content (Network City. Education; Moscow Electronic School; Russian electronic School; Yandex.School);
- Information resources that provide ready-made recordings of video lessons, lesson scenarios or fragments thereof (Educotion.yandex.ru; Kopilkauro-koy.ru);
- Information resources that provide the opportunity to perform tasks under the supervision of a teacher (Uchi.ru; Resh.edu.ru);
- Services that provide the ability to solve tasks of the state final certification and homework (Gdz.ru; Sdamgia.ru; 4ege.ru);
- Integrated learning systems with structured content and the possibility of an individual approach to student learning (SkySmart.ru; Algoritmica.org; Physics.ru).

The lack of implemented Russian developments and the imperfection of most of the specialized information resources used during the pandemic were the reasons for the activation of real work in the direction of designing and developing intelligent learning systems.

Thus, adaptive systems and e-learning environments based on artificial intelligence have become an actual technology for monitoring and evaluating knowledge. To what extent did these systems ensure the quality of subject education and the level of objectivity of students' assessment activities during the pandemic? How can we evaluate the effectiveness and efficiency of the process of pedagogical control and assessment of students' knowledge based on artificial intelligence methods?

The formulated questions determined the problem of this study.

3 Survey Methodology

Fully adaptable electronic learning systems based on artificial intelligence have become an actual methodology during the unplanned transition to the distance learning format. The functionality of the modern ITS is represented by a complex distributed system, the main components of the architecture of which are an intelligent interface, a learner model, a domain model, an adaptive learning model, a knowledge and competence assessment model, and a repository of educational objects.

The flexible learner model is an innovative approach to education that focuses on the development of learner competences. It is based on the results of diagnostics of the learner's key characteristics, such as personal qualities, motives, initial and target competences, as well as his/her individual learning style. The formation of the learner model allows taking into account the learner's initial level of knowledge and motivational readiness for learning. This means that the learning process is built taking into account the individual characteristics of each student, which makes it more effective and efficient. In addition, the flexible learner model allows predicting the results of pedagogical influence, determining what competences the learner will be able to master as a result of training. This allows for the development of an individualised learning plan aimed at achieving specific goals. The flexible learner model also provides variability in the presentation of learning material, which makes the learning process more interesting and entertaining for students. In addition, it allows step-by-step tracking of learning progress, which makes the learning process more transparent and controllable. In general, the flexible learner model is a modern tool in education that allows creating a more efficient, individualised and effective learning system. It contributes to the development of learners' competences and helps them to achieve success in learning [1].

The core of ITS architecture is an adaptive learning model that implements a sequence of control actions in relation to each of the subsystems and the interpretation of external user influences. The learning model includes a set of models of trainees, a set of learning strategies, as well as the function of generating educational scenarios. Its functioning is based on the principle of cyclic monitoring of the state of the student's model by means of intermediate tests after passing the next module of the course. The technological level of modern IOS makes it possible to expand the possibilities of intellectualization of the educational process due to the ability of the system to independently solve the tasks that it should teach the student to solve, which significantly simplifies the procedure for developing a bank of educational tasks, provides a deeper analysis of the student's responses and better quality counseling.

4 Data and Methods

From September 2019 to May 2021, the authors conducted a pilot study with the participation of 711 schoolchildren of grades 10–11 studying a course of mathematics in a mixed format (including during the pandemic period completely in a distance format), as well as with the inclusion of an adaptive intellectual learning component throughout the entire training period. The schools of the Lipetsk region were the experimental base of the study. In control classes ($n1 = 155 + 140 + 151 = 446$) classes were conducted using mixed learning models, including information resources and platforms (Russian Electronic School; Yandex.School; Uchi.ru; Yaklass.ru; Infourok.ru; Resh.edu.ru). Training in experimental classes ($n2 = 86 + 82 + 97 = 265$) was also built in a mixed format, but on the basis of a hybrid intellectual learning system with an adapted procedure for providing educational materials that provide an intellectual assessment of the mastery of subject knowledge, competencies and procedures.

The model developed by the authors [2] and the hybrid intelligent system introduced into the practice of teaching and accompanying the student's educational and research activities in mastering the mathematics of complex knowledge differs from other "adaptive" models in that, firstly, the use of a hybrid paradigm in the design of intelligent systems allowed to minimize risks errors in tracing individual trajectories. The proposed architecture of an adaptive learning system with an inbuilt knowledge assessment module based on the fractal approach improves the quality of learning and assessment using artificial intelligence [3, 4]. The system is oriented not only to the acquisition of subject knowledge and skills, but also to the development of the learner's personality in the process of research activities using intelligent control. Fuzzy sets, fractal geometry, complex numbers, random events and quantities are offered as examples of complex mathematical knowledge for the organisation of teaching and research activities. Such pedagogical support, including clusters of research tasks detailed according to the levels of students' scientific potential, is a novelty of this approach. It should be noted that the current model of the unified state exam in mathematics has not only sufficient diagnostic potential, but also served as the basis for the development of a promising model of the profile unified state exam, structurally and meaningfully modified [5, 6]. Already in 2022, the content of control and measuring materials includes tasks for the construction of probabilistic models of real situations, tasks with complex numbers [7–9].

The effectiveness of the hybrid intellectual system of teaching mathematics was evaluated based on the results of the Unified State Exam in the 11th grade, which, as mentioned above, is the main form of state assessment of knowledge and competencies in the Russian Federation. As a criterion for evaluating the effectiveness of teaching mathematics using a hybrid intelligent system, complex tasks with a detailed answer (profile level) were selected. The model of the profile Unified State Exam has not changed for three years [10–12]. Data analysis is based on the following rule: for the correct completion of the first three tasks (13–15) with a detailed answer, two points are awarded, for the next two (16, 17)—3 points for each. The maximum possible points are given by the last two tasks (18, 19)—for each of them you can get 4 points. The answers are checked by experts on the basis of an approved work evaluation system. Tasks evaluate the assimilation of educational material in the modules: "Algebra and the beginning of mathematical analysis" (13, 15, 18 and 19), "Geometry" (14, 16) and "Practice-oriented tasks" (17). In these tasks, emphasis is placed on checking the mastery of the algebraic apparatus; on checking the development of basic ideas of mathematical analysis; on assessing the formation of geometric representations, the ability to analyze a geometric design; on assessing the ability to find a solution to a problem of increased and high complexity [13–15].

5 Results and Discussion

The hypothesis of the study on the effectiveness of teaching mathematics using a hybrid intelligent system and the accuracy of evaluating the educational results of students based on the structure of an artificial neural network was experimentally confirmed. Table 1 presents the data of comparative diagnostics of the content sections of complex tasks of the Unified State Exam in Mathematics (profile level) in the control and experimental groups.

As follows from the set of individual results of the state evaluation procedure, there is a slight difference between the sample average values of the primary score

Table 1 Statistical analysis of the performance of tasks with a detailed response in dynamics over 3 years using descriptive statistics methods (sample average)

Classes	Average primary score		
	2019	2020	2021
Control	2,49	2,21	2,59
Experimental	2,67	3,30	3,03

scored in the control and experimental classes. There is a weak positive dynamics of the change in this indicator in the period from 2019 to 2021, which is explained by the gradual adaptability of participants in the educational process to the new format of blended learning during the coronavirus pandemic and the integration of innovative electronic tools, including intellectual systems into the system of mathematical education. It should be noted that since the spring of 2020, the training format has been distance learning, 2021—mixed. Statistical verification using the Student's t-test also confirmed this assumption (Table 2).

In connection with the transition to a mixed format of education in 2021, the average value of the primary score scored in the control classes slightly increased and, as a result, the empirical value of the criterion decreased. At the same time, such dynamics does not allow us to talk about a cardinal increase in the level of training of students using hybrid intelligent systems.

Figures 1–3 show the distribution of the results of math tasks with detailed answers between control and experimental groups. There is a smooth annual increase in the number of high scores in experimental groups associated with in-depth study of the subject within the framework of the student's educational and research activities for the development of complex mathematical knowledge using a hybrid intellectual system. Statistical verification using Pearson's χ2–criterion established significant differences between experimental and control groups: 2019 $-\chi^2 = 55{,}40$, by $p < 1{,}52*10^{-6}$; 2020 year $-\chi^2 = 91{,}34$, by $p < 1{,}41*10^{-12}$; 2019 year $-\chi^2 = 97{,}65$, by $p < 9{,}54*10^{-14}$.

The share of participants with low scores in 2019 is sharply decreasing compared to 2020. The group of participants who scored high scores has not yet undergone significant changes.

An analysis of the results of math assignments with detailed answers in 2021 showed that no more than 10% of schoolchildren score high (compared to last year, the indicator increased by 5%). Thus, a certain increase in results at the level of high scores is noticeable.

Table 2 Statistical analysis of the completion of tasks with a detailed response over 3 years using the Student's t-test

Statistics/Probability	2019	2020	2021
t	0,189	1,948	1,274
p	< 0,28	< 0,03	< 0,10

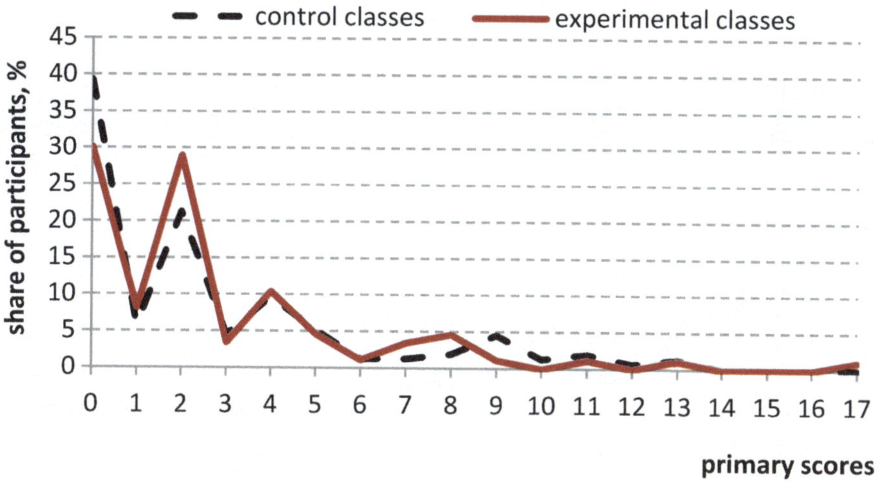

Fig. 1 Comparison of the distribution of the results of math tasks with the detailed answers of control and experimental groups in 2019

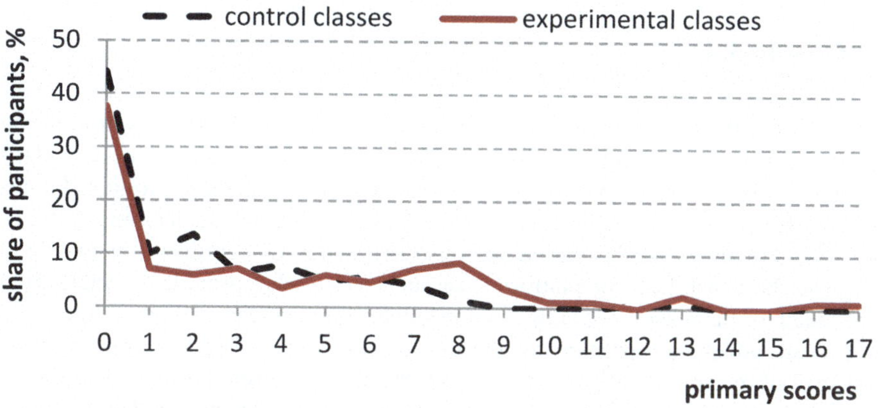

Fig. 2 Comparison of the distribution of results of math tasks with detailed answers of control and experimental groups in 2020

Against the general background of the decline in the results of the unified state exam related to the Covid-19 pandemic and the transfer of training to distance learning, the presented empirical data still demonstrate a slight positive trend in contrast to the neighboring regions, for example, in the Voronezh Region, negative dynamics is recorded [5, 16, 17].

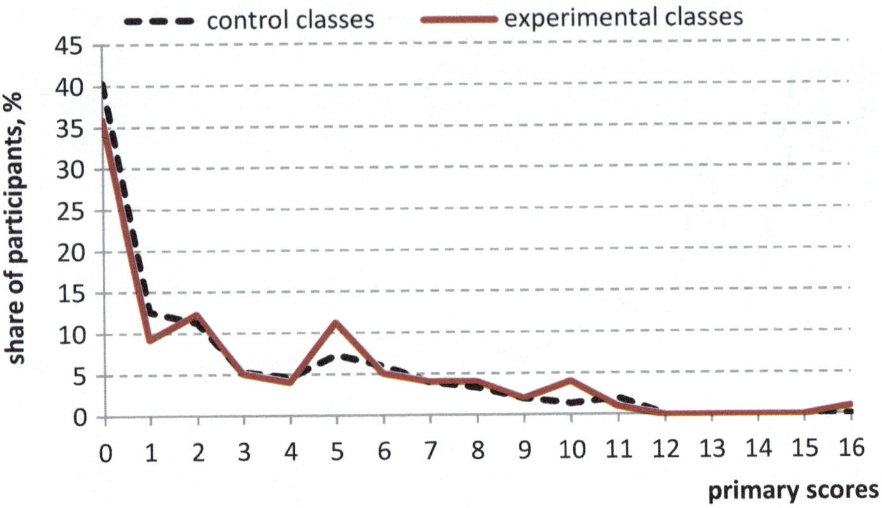

Fig. 3 Comparison of the distribution of the results of math tasks with the detailed answers of control and experimental groups in 2021

6 Conclusion

Summarizing the results obtained, it can be concluded that the preservation of the quality of Russian education in the context of the global crisis of the Covid-19 pandemic was ensured both through the widespread use of modern station technologies and, to a greater extent, through the introduction of intellectual learning systems. The forced regime of self-isolation demonstrated that not all graduates were able to fully use the capabilities of platforms, online services that provide an intellectual assessment of the mastery of subject knowledge, competencies and procedures, and as a result, adequately prepare for the unified state exam. The key point was the low level of discipline and activity of modern schoolchildren. Distance learning is a form of education for motivated students who are able to independently master mathematical knowledge, to solve complex problems (tasks) with a lack of initial data in a situation of ambiguity and uncertainty, to search and adapt to rapid changes in the environment. Adaptive learning systems based on a combination of intelligent methods play an important role in this aspect.

Such systems are able to solve a binary problem: on the one hand, they provide the possibility of continuous diagnostics of the educational process – tracking the student's progress in achieving the specified results, determining the current state of learning and fixing the required level for successful final certification. On the other hand, intellectual learning systems are focused on teaching subject knowledge and skills, on the effective development of the learner's personality through the implementation of the intellectual management functionality in the "teacher—computer—learner" triad, which became especially important during the period of forced self-isolation. The hybrid intellectual system proposed by the authors to support the student's design and research activities in mastering the mathematics of complex knowledge successfully solves the tasks set. Moreover, it has proven its effectiveness in the conditions of distance and mixed learning. Increasing the success of solving complex tasks at the profile level is possible when solving problems that require "disclosure of the essence of complex mathematical knowledge", the development of critical thinking and intuition are included in the learning process. This requires shifting the emphasis in teaching mathematics in high school from memorizing definitions and solving a large number of technical numerical problems to solving research problems. The hierarchical system of research practice-oriented tasks, differentiated according to the levels of complexity of the elements of the generalized construct of mathematical knowledge, constitutes a multi-level model of the subject area in the hybrid intellectual environment implemented by the authors.

Acknowledgements The reported study was funded by RFBR, project number 19-29-14009.

References

1. Johanes P, Lagerstrom L (2017) Adaptive Learning: The Premise, Promise, and Pitfalls. Paper presented at the 124th ASEE Annual Conference and Exposition. Columbus, Ohio, p 27538
2. Dvoryatkina S, Khizhnyak A, Smirnov E, Shcherbatykh S, Zhuk LV (2021) Open innovation model of student's research activities (Running title: Open Innovation Model). J of Teach Ed for Sustainability 23(2):82–95
3. Smirnov E, Dvoryatkina S, Shcherbatykh S (2020) Parameters and structure of neural network databases for assessment of learning outcomes. Int J of Criminol and Sociol 9(83):1638–1648
4. Dvoryatkina S, Smirnov E, Lopukhin A (2017) New opportunities of computer assessment of knowledge based on fractal modeling. Paper presented at the 3th International Conference on Higher Education Advances, HEAd'17. Universitat Politecnica de Valencia, Valensia, pp 854–864
5. Daines J, Troka T, Santiago J (2016) Improving Performance in Trigonometry and Pre-Calculus by Incorporating Adaptive Learning Technology into Blended Models on Campus. Paper presented at the 123rd Annual ASEE Conference & Exposition. New Orleans, Louisiana, p 25624
6. Bowen WG, Chingos MM, Lack KA, Nygren TI (2014) Interactive learning online at public universities: evidence from a six-campus randomized trial. J of Policy Anal and Manag 33(1):94–111
7. Brown M, McCormack M, Reeves J, Brook DC, Grajek S, Alexander B, Gannon K (2020) 2020 Educause Horizon Report, Teaching and Learning Edition. EDUCAUSE, Louisville, CO
8. Brusilovsky P (1996) Methods and techniques of adaptive hypermedia. Adaptive hypertext hypermedia 6(2–3):87–129
9. Chudinsky RM, Malev VV, Mosolov ON, Dubov VM (2021) Analysis of the results of the unified state exam in the Voronezh region in 2020. World of Sci Pedag and psychol 2(9)
10. Joo J, Spies RR (2019) Aligning Many Campuses and Instructors around a Common Adaptive Learning Courseware in Introductory Statistics. Lessons from a Multi-Year Pilot in Maryland
11. Natriello G (2017) The adaptive learning landscape. Teach Coll Rec 119(3):1–46
12. Prusty BG, Russell C (2011) Engaging students in learning threshold concepts in engineering mechanics: adaptive eLearning tutorials. Paper presented at the 17th International Conference on Engineering Education (ICEE), pp 21–26

13. Smirnov E, Dvoryatkina S, Martyushev N, Shcherbatykh S (2023) Software package to support students' research activities in the hybrid intellectual environment of mathematics teaching. Mathematics 11(4):952
14. Shutaleva A, Starostin A, Nikonova Z, Savchenko I (2022) Migration potential of students and development of human capital. Educ Sci 12(5):324
15. Tsarev RY, Tynchenko SV, Gritsenko SN (2016) Adaptive learning using the resources of the information and educational environment. Modern Probl of Sci and Ed 5:219–219
16. Martyushev N, Shutaleva A, Malushko E, Nikonova Z, Savchenko I (2021) Online communication tools in teaching foreign languages for education sustainability. Sustain (Switz) 13(19):11127
17. White G (2020) Adaptive learning technology relationship with student learning outcomes. J of Inform Technol Ed: Res 19:113–130

Visualization in Blended Learning of Geometry in the High School

Valentina Manyakhina, Polina Makarova, and Irina Smirnova

Abstract During the COVID-19 pandemic, many teachers reconsidered their attitudes to-ward online learning. According to surveys, they will combine online educational materials and face-to-face classroom practices. Blended learning in mathematics education provides personalized learning, alternative methods for teacher-student feedback and communication. In addition, mathematics teachers get the opportunity to make extensive use of 3D graphics technologies, AR applications, dynamic geometry environments for visualization. Our goal is an empirical study of the conditions for improving the effectiveness of teaching stereometry in high school through the use of visualization technologies and blended learning. Our re-search confirmed that after forced distance learning in the COVID-19 pandemic there has been a positive change in the attitude of math teachers towards blended learning, and that high school students are ready for blended learning. A blended learning model for geometry was proposed, a feature of which is that the content of the digital learning environment is based on the principles of visualization, and e-tools for visualization are actively used in teaching. The conducted research proved the success of the proposed model.

Keywords Blended learning · Mathematics education · Geometry in high school · Visualizations in education

V. Manyakhina (✉) · P. Makarova · I. Smirnova
Institute of Mathematics and Informatics, Moscow Pedagogical State University (MPGU), 14, Krasnoprudnaya St, 119991 Moscow, Russia
e-mail: vg.manyakhina@mpgu.su

P. Makarova
e-mail: ps.makarova@mpgu.su

I. Smirnova
e-mail: im.smirnova@mpgu.su

A. L. Semenov et al. (eds.), *Structural and Technological Transformation of Education in the Post-Pandemic Period*, Lecture Notes in Networks and Systems 1282,
https://doi.org/10.1007/978-3-031-84039-5_4

1 Introduction

In recent years, there has been a large-scale digital transformation of Russian education, actively forming a digital educational environment. State and non-state educational online platforms for students have been created, such as "Russian E-School", "Moscow E-School", "Ya-Klass", "Uchy.ru" and others, which provide teachers with all the opportunities for e-learning and blended learning.

The digital transformation of education is aimed at improving educational outcomes. Blended learning, the essence of which is a combination of face-to-face and online instruction, is one of the main trends of modern education. Discussions about the approach of blended learning have long been held by experts, teachers, and school administrators. Now there is no doubt about the need for blended learning in schools [1, 2]. Blended learning approach provides an opportunity to implement a fundamental shift in the existing model of learning towards personalized learning [3]. In addition, we see other benefits of using blended learning models in high school mathematics classes—digital technology makes the use of visualization technologies in teaching more accessible and opens up new opportunities for visualization in geometry and stereometry classes [4], and most importantly, the digital environment creates conditions for individual work of students with geometric 2D and 3D models [5–7].

The pervasive dominance of visual information through the widespread use of all kinds of digital tools and technologies is observed, and the influence of the "visual turn" on the children of Generation Z and ALPHA certainly cannot be ignored. Nowadays children and adults receive the majority of information visually, so it is undoubtedly important and necessary to use visualization technologies, especially when it comes to developing learning resources for digital learning environments.

The very notion of visualization is defined by us as a general method of presenting the studied information or phenomenon in a convenient and observable form for further analysis. Visualization is one of the directions of such a complex concept as "visual thinking".

The intentional and deliberate use of visualization can improve the effectiveness of teaching mathematics (especially geometry and stereometry) [8–10]. Speaking about visualization, we mean not only teaching with many visual aids, but also active work of students with visual geometric models, some of which can be made by students themselves. In this case, it is possible to avoid several problems related to the lack of visual support of abstract scientific concepts and the detachment of analytical processes from visual objects. Therefore, we propose to implement such visual support through e-tools in blended learning environment. Our goal is an empirical study of the conditions for improving the effectiveness of teaching stereometry in high school using visualization technologies and blended learning.

This study is aimed at receiving answers to the following questions:

1. How did distance learning in 2020 during the COVID-19 pandemic affect attitude of students and teachers toward blended learning. Are students willing to learn partially online as part of blended learning? Are math teachers willing to continue

working in an online environment alongside traditional face-to-face instruction? Furthermore, which blended learning models are most appropriate for the current environment?

2. Would high school blended learning of stereometry be more effective than traditional learning if the course content was designed to be visualized in a blended learning environment?

2 Survey Methodology

This study used a mixed methods design that combines both quantitative and qualitative approaches. Two phases of the study were conducted.

Phase 1: A study of the attitude of teachers and students to blended learning after forced remote learning during the COVID-19 pandemic in 2020.

Phase 2: Research on the effectiveness of blended learning of stereometry, provided that the digital resources of the online course were developed using visual technologies.

2.1 Survey Setting

Phase 1. Survey among teacher and students.

For the research purposes an online questionnaire for teachers was administrated in 2019 (before the COVID-19 pandemic), and in 2021. Mathematics teachers who work in schools in Moscow participated in the survey. Both surveys were carried out to clarify how exactly the mathematics teachers apply the digital resources of online educational platforms, which blended-learning models they already use or are ready to use in the future.

In 2021, an online survey was conducted among Moscow school students, who were forced to switch to distance learning in 2021. The questionnaire was administered to identify the problems of online mathematics instruction and attitudes toward the possibility of switching to blended learning in the future.

Phase 2. Blended learning of stereometry for high school students using visualization technologies. Among the experimental models, the pretest–posttest control group model was used in the study.

Participants. Forty-three 10th grade male and female students from School No 949 in Moscow were selected to achieve the goals of this study. The selection was carried out according to the results of the state exam in mathematics (grade 9), based on the fulfillment of tasks in geometry.

To participate in the experiment, 43 students from two 10th grades who scored between one standard deviation above and below the mean were selected. Two groups of students were surveyed: a group of students ($n = 19$) from one 10th grade, marked as experimental and a group of students ($n = 24$) from another grade as control group.

Note that students in grade 10-IT with high levels of digital skills were selected for the experimental group. The groups were taught differently, the experimental group used a blended learning of stereometry, while the control group used the traditional method of teaching stereometry.

Model and environment of blended learning of stereometry. The results of the previous survey of teachers and students were considered, and the experimental teaching on a combination of two models of blended learning was built—"Flipped Classroom" and "Laboratory Rotation" [11]. A digital learning environment is a prerequisite for blended learning [12].

Teachers' feedback to students about their learning progress was essential [13] and was also carried out in this environment (messaging in chats and forums, feedback on individual tasks).

Visualization e-tools. Geogebra has become an important visualization tool [14]. The teacher created many 2D and 3D models in Geogebra and the students also worked systematically in this environment. Moreover, the online whiteboard Miro was used, Mindmeister for custom projects, brainstorming, and mind mapping, interactive worksheets Wizer.me.

Digital content was created for the online course and we relied on the following principles in developing it:

1. In the design of drawings and solutions of problems we introduced color differentiation, the solutions were presented in the form of storyboards. When constructing sections of polyhedrons, each stage is marked with a different color. Meaning of color (Fig. 1):

2. To systematize and remember the theory better, mindmaps with color differentiation were made in the interactive online whiteboard Miro. For example, in the mindmap about mutual arrangement of points, lines and planes in space each branch has its own color and value.
3. Obligatory individual research work of students in GeoGebra 3D Calculator, both with ready-made models and constructed independently [15, 16].

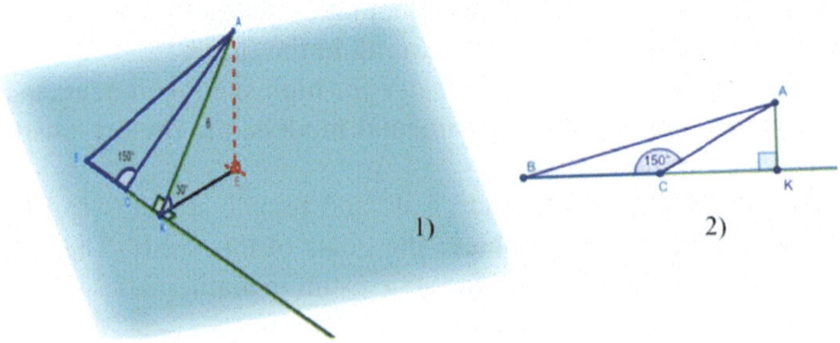

Fig. 1 Color differentiation: Blue—elements that are given in the problem condition; Green—elements to be added; Red—elements to be found at this stage. Instead of color you can use all kinds of dashed, wavy, jagged lines

The blended learning model we used was based on a combination of two models: "Flipped Class" and "Lab Rotation". The principle of switching models was as follows: if a simple topic was studied, then the "Flipped Class" model was applied, when a complex topic was studied, then the "Lab Rotation" model was applied and the teacher conducted the lesson in the computer lab, therefore, the complicated theoretical material was first analyzed together with the teacher and after students performed individual tasks on the computer, working with the online course.

2.2 Data and Methods

Data collection procedure of phase 1.

The sudden shift to remote learning during the pandemic had a significant impact on the use of educational technology in schools and provided extensive material for researchers. Many researchers used surveys of teachers, students, and parents to understand how they adapted to remote learning and their attitudes toward distance learning [17–19].

Two online surveys of mathematics teachers were conducted: in 2019 (before the COVID-19 pandemic) and in 2021. Fifty-three mathematics teachers ($n = 53$) who work in Moscow schools took part in the survey. More than half of the respondents (56%) were young teachers who had recently graduated from Moscow State Pedagogical University. The sample of schools is random, they are not characterized by particular qualities and can be considered average Moscow educational institutions.

In 2021, we conducted an online survey of Moscow school students ($n = 120$), who were forced to switch to distance learning. These were students in grades 10 and 11 who study at the same schools where the participating teachers are employed.

Data collection procedure of phase 2.

As an entry test (pretest), we took the results the main state examination (MSE) of the 9th grade. The MSE includes 8 geometry assignments (Nos. 15–19 and 23–25), which are scored a total of 32 points. For 5 months, students of the experimental group studied stereometry in the above-described blended learning environment.

To organize the final assessment (posttest), we used the Statgrad final diagnostic work for the 10th grade in mathematics in the format of the Unified State Exam (USE). The Statgrad system is a system for preparing students for the State final certification in all subjects, conducted by the Moscow Institute of Open Education and the Moscow Center for Continuous Mathematics Education. The Unified State Exam in Mathematics includes 5 geometry tasks, 2 of which are for stereometry. The performance of these tasks by students in the experimental and control groups were considered as a result of exit testing. One could get a total of 32 points for the correct completion of all 5 tasks.

To determine the reliability of the coincidence (before the experiment) and the reliability of the difference (after the experiment) of the characteristics of the compared samples (control and experimental groups), we calculate the values of the Kramer-Welch criterion using the formula:

$$T = \frac{\sqrt{n_1 \bullet n_2} \bullet |\overline{x_1} - \overline{x_2}|}{\sqrt{n_1 \bullet S_1^2 + n_2 \bullet S_2^2}}.$$

3 Results and Discussion

The results obtained are presented separately for phase 1 (a study of the attitude of teachers and students to blended learning after forced remote learning during the COVID-19 pandemic in 2020) and phase 2 (research on the effectiveness of blended learning of stereometry, provided that the digital resources of the online course were developed using visual technologies).

3.1 How Isolation and Lockdown Have Affected Blended Learning

The results of the questionnaire in 2019 showed that only 14% of the teachers surveyed did not use digital resources of online educational platforms at all, 86% used them, and more than half of teachers did it at every lesson or quite often (Table 1).

The emergency transition to a distance learning format associated with the COVID-19 pandemic required all participants in educational relations to adapt to the new conditions of learning in a stressful mode. Despite distance learning having triggered mixed reactions in the society—parents, learners, and even teachers, this situation has stimulated process of introduction of digital technologies in education. In the spring of 2021, we re-interviewed the same teachers. Despite the difficulties, most of the teachers (74%) positively evaluated the experience of forced distance learning but noted that they were partially satisfied with the quality of distance learning, a significant increase in workload, and technical problems. Dissatisfaction with the quality of distance learning largely influenced the teachers' answers to the question "What form of education would you prefer in the future, if you had a choice?" only 6% of the teachers surveyed would choose distance learning. It is important to note that by actively using distance learning technologies, many teachers mastered them, and grew to appreciate their didactic capabilities, so the majority of teachers 57% spoke in favor of combining traditional full-time education with elements of distance learning, that is, for blended learning.

The research conducted by the Laboratory of Media Communications in Education of the National Research University Higher School of Economics in 2020 also showed that on the national scale the share of teachers who began to use digital online resources in their work increased from 64 to 85%. At the same time, three-fourths of teachers, who had not previously used any online digital educational platforms, began to use them. Of these, 47% noted that they are likely to continue using

Table 1 Results of a survey of mathematics teachers

Question	2019 *n* (%)	2021 *n* (%)
How often do you use online educational platforms in the classroom?		
I don't use	8 (15%)	2 (3%)
Rarely	22 (41%)	14 (27%)
Often	19 (36%)	30 (56%)
In every lesson	4 (8%)	7 (14%)
What blended learning models do you use?		
I do not	31 (58%)	20 (36%)
"Flipped classroom"	14 (27%)	21 (40%)
"Laboratory rotation"	5 (9%)	6 (12%)
"Zone rotation"	3 (6%)	3 (6%)
Other	0 (0%)	3 (6%)
From your point of view, what are the benefits of blended learning? (Give up to three reasons)		
Better organization of the lesson	25 (47%)	31 (58%)
Better organization of students' independent work	21 (40%)	33 (62%)
Higher motivation of students	36 (68%)	30 (56%)
Greater activation of learning activity	31 (59%)	36 (68%)
New forms of organizing learning activities	12 (22%)	10 (20%)
Personalization	10 (18%)	13 (25%)
Other	5 (9%)	6 (12%)
How do you rate the distance learning experience during the COVID-2019 pandemic?		
Positive	–	39 (74%)
Negative	–	13 (24%)
Difficult to answer	–	1 (2%)
Are you satisfied with the quality of distance learning?		
Yes	–	6 (11%)
Partially	–	22 (41%)
No	–	22 (41%)
Difficult to answer	–	3 (7%)

(continued)

Table 1 (continued)

Question	2019 *n* (%)	2021 *n* (%)
In your opinion, how did the teachers' workload change during the distance learning period?		
Remained the same	–	9 (17%)
Decreased	–	0 (0%)
Increased	–	39 (74%)
No answer	–	5 (9%)
What form of education would you choose in the future if you had a choice?		
Distance education	2 (4%)	4 (6%)
Traditional education	32 (60%)	17 (33%)
Blended learning	18 (34%)	30 (57%)
Difficult to answer	1 (2%)	2 (4%)

them in their work in the future. The results of our survey of Moscow teachers in 2021 correlate with all-Russian data—only 3% of teachers do not use digital online platforms, preferring traditional face-to-face instruction (in 2019 was 15%). This certainly creates the conditions for a more active implementation of blended learning approach in schools, both in Moscow and in Russia as a whole.

However, the results of the questionnaire in 2021 showed that more active use of digital educational resources by teachers practically does not change the traditional forms and methods of teaching. Responding to the question "What blended learning models do you use?"—40% of teachers indicated that they occasionally use the "Flipped Classroom" model as the easiest blended learning model in terms of implementation. Quite a few teachers (34%) indicated that they do not use any blended learning models but use the resources of the digital learning environment as visual material when explaining a new topic, as well as for testing students and for performing practical tasks with automated testing. That is, teachers mainly use the capabilities of digital educational platforms to optimize their work when preparing and conducting a lesson (for example, searching for a ready-made lesson script from the library of the "Moscow E-School" and broadcasting the script to an interactive panel or the board during the lesson), and to optimize the checking of the tasks completed by students (automated checking of tests and interactive tasks).

Although more than half of the teachers we surveyed do not use blended learning models, they understand its benefits. The most popular answers: "blended learning leads to an intensification of learning activities", "contributes to an increase in learning motivation", a more effective organization of the lesson was noted as well.

Survey among students. Answering the question "How have you adapted to the new conditions of online distance learning?", about 8% of students said that they poorly adapted to the new distance learning environment, 23% of students assessed the level of adaptation as "satisfactory", and almost 70% of students—as "good" and "excellent" (Table 2).

Table 2 Results of a survey of students

Question	2021 *n* (%)
How have you adapted to the new online distance learning environment? Excellent Good Satisfactory Poor, practically did not adapt	30 (25%) 53 (44%) 27 (23%) 10 (8%)
Is it convenient for you to study online? Yes, it is convenient Yes, but difficult No, very difficult No, too easy Difficult to answer	30 (25%) 49 (41%) 31 (26%) 1 (1%) 9 (7%)
What difficulties have you encountered in online learning? (specify no more than 3 problems) Technical problems Lack of proficiency in IT Difficulty working in a digital environment Difficulty completing assignments without teacher's guidance Lack of face-to-face interaction with teacher Other	81 (68%) 31 (26%) 60 (50%) 53 (44%) 88 (74%) 21 (9%)
Are you satisfied with the quality of distance learning? Yes No Partly Difficult to answer	11 (9%) 45 (38%) 59 (49%) 5 (4%)
What form of education would you choose in the future, if you had a choice? Distance education Traditional education Blended learning Difficult to answer	2 (2%) 18 (15%) 94 (78%) 6 (5%)

In general, the difficulties faced by students during distance learning are predictable—these are technical problems (68%), the difficulty of working in an online environment (50%). Many high school students experience difficulties in learning, in completing assignments, if there is no face-to-face contact with the teacher. Moreover, it is necessary to pay attention to the fact that 26% of the students pointed out insufficient knowledge of IT (26%), which prevents them from switching to blended learning effectively. To the question "Is it convenient for you to study online?", only 66% of the surveyed students responded that it is convenient, but of these 41% of students indicated that they experience difficulties in distance learning. Obviously, this circumstance also affected the assessment of the quality of distance learning. Only 9% are completely satisfied with the quality of distance learning, the rest are either partially satisfied (49%) or completely dissatisfied with the quality of distance learning. However, after forced distance learning during the COVID-19 pandemic, we can observe that high school students have become more prepared for blended learning, as evidenced by the results of the survey—78% of students expressed their readiness for blended learning and only 15% of students are against blended learning and vote for the traditional learning format.

4 Conclusion

After being forced into distance learning during the COVID-19 pandemic, teachers' willingness to use blended learning has increased. However, it must be recognized that the integration of digital technologies into the educational process has not yet led to its radical transformation. According to the SAMR (The Substitution-Augmentation-Modification-Redefinition Model), the use of digital technologies in Russian educational institutions, particularly in Moscow schools, so far leads only to the replacement or improvement of traditional teaching tools but does not lead to a change and radical transformation of teaching practice. Mathematics teachers point to the lack of personal digital devices for each student in the classroom, or at home. Of course, it should be considered that in the majority of schools personal computers and laptops for students are present only in computer classes, in common classrooms there is only a computer for the teacher. A "LabRotation" model of blended learning can be used as a remedy for the problem, however, this is not always possible, as computer science lessons take place in a computer lab. Another option is to ask students to bring their own personal tablets or laptops (BYOD) to class, but many parents do not support this initiative. The "Flipped-Classroom" model is designed to have a personal device in the student's home. Apparently, right now, it is the easiest and the most effective blended learning model to implement.

In general, high school students have adapted well to online learning and are willing to do some of their academic work in an online environment in the future. Moreover, many high school students ask teachers to conduct additional classes remotely (including preparation for the USE). It can be concluded that high school students are ready for blended learning.

The conducted experimental blended learning of stereometry using "Flipped-Classroom" and "LabRotation" models demonstrated that these models are relatively and quite effective in high school provided that the teacher and students have a high level of digital skills. However, the teacher needs to control students' preparation for the lesson if the "Flipped-Classroom" model is used, otherwise the work in the lesson will be ineffective.

Based on the findings of the present study, it can be argued that blended learning of geometry in high school creates conditions for the active use in teaching of visualization technologies, digital tools for constructing and studying the properties of 3D and 2D models. This contributes to the development of spatial representation in students, helps them to better understand theory and solve problems in geometry. Students' grades in the experimental group were significantly higher than the grades obtained by the students in the control group with a traditional setting. The students of the experimental group expressed a generally positive opinion about the resources of the online course and the blended instruction method, which gave them the opportunity to work autonomously at their own pace, while the ability to interact with 3D and 2D models was extremely useful for visualizing stereometric figures.

This study should be seen as an initial approach and supplemented to understand which visualization technologies contribute to a deeper understanding of the high school stereometry course.

References

1. Dziuban CD, Picciano AG, Graham CR, Moskal PD (2015) Conducting Research in Online and Blended Learning Environments: New Pedagogical Frontiers, 1st edn. Routledge
2. Lin Y-W, Tseng C-L, Chiang P-J (2017) The effect of blended learning in mathematics course. Eurasia J Math Sci Technol Educ 13(3):741–770
3. Da Silva NM, Capuchinho AC (2017) Blended learning: concepts, reflections and possibilities of learning personalization. CEUR Work Proc 1877:148–156
4. Kadunz G, Yerushalmy M (2015) Visualization in the teaching and learning of mathematics. Paper presented at the 12th International Congress on Mathematical Education, Cho S (ed). Springer, Cham
5. Engelbrecht J, Llinares S, Borba MC (2020) Transformation of the mathematics classroom with the internet. ZDM Math Ed 52:825–841
6. Fernandes L, Lanzellotti L, Lopes C, Cunha GG, Landau L (2015) Augmented reality to support 3D visualization of geometric elements in classroom with AR teacher tools. Paper presented at the 14th International Conference on Modeling and Applied Simulation, MAS 2015, pp 43–48
7. Hall J, Chamblee G (2013) Teaching algebra and geometry with Geogebra: preparing pre-service teachers for middle grades/secondary mathematics classrooms. Comp in the Schools 30(1–2):12–29
8. David MM, Tomaz VS (2012) The role of visual representations for structuring classroom mathematical activity. Ed Stud in Math 80(3):413–431
9. Kadunz G (2000) Visualization, image and metaphor: the mediating activity of visualization in learning mathematics. J for Math Didact 21(3–4):280–302
10. Presmeg N (2014) Contemplating visualization as an epistemological learning tool in mathematics. ZDM Int J on Math Ed 46(1):151–157
11. Graham CR (2009) Blended learning models. In: Encyclopedia of Information Science and Technology, 2nd edn. pp 375–382
12. Graham CR, Borup J, Pulham E, Larsen R (2019) K–12 Blended teaching readiness: model and instrument development. J of Res on Technol in Ed 51(2):1–18
13. Boud D (2015) Feedback: ensuring it leads to enhanced learning. Clin Teach 12:3–7
14. Kramarenko TH, Pylypenko OS, Muzyka IO (2020) Application of Geogebra in stereometry teaching. CEUR Work Proc 2643:705–718
15. Kusumah YS, Kustiawati D, Herman T (2020) The effect of Geogebra in three-dimensional geometry learning on students' mathematical communication ability. Int J of Instruct 13(2):895–908
16. Saha RA, Ayub AFM, Tarmizi RA (2010) The effects of Geogebra on mathematics achievement: enlightening coordinate geometry learning. Proc—Soc and Behav Sci 8:686–693
17. Bubb S, Jones M-A (2020) Learning from the COVID-19 home-schooling experience: listening to pupils, parents/carers and teachers. Improv Schools 23(3):209–222
18. Huber SG, Helm C (2020) COVID-19 and schooling: evaluation, assessment and accountability in times of crises–reacting quickly to explore key issues for policy, practice and research with the school barometer. Ed Assess Eval Account 32(2):237–270
19. König J, Jäger-Biela DJ, Glutsch N (2020) Adapting to online teaching during COVID-19 school closure: teacher education and teacher competence effects among early career teachers in Germany. Eur J of Teach Ed 43(4):608–622

Popular Science Film as a Resource for Integrating Media Education Technologies into the Mathematics Teaching System

Svetlana Dvoryatkina, Angelica Dyakina, and Tatiana Safronova

Abstract The formation of a media–literate generation is one of the priority directions of Russia's social development. It is possible through the systematic integration of media education technologies into an integrated educational system at all levels. In this study, the technology of media education has been developed and implemented in the process of training future teachers of mathematics. Media texts in the format of popular science films are considered as effective didactic means of forming stable motives for research and mastering complex mathematical knowledge. Future mathematics teachers, who will teach the digital generation, should master media educational technologies at the stage of university training and learn how to implement them into the taught subject. "Integrated media education" is considered by the authors as a cultural and pedagogical phenomenon based on the creative activity of the teacher. Of paramount importance in it is the processality of learning with the actualization of critical thinking, creative and communicative abilities of the individual. The features of the integration of media education into individual disciplines are associated with a new means of teaching–educational media text, made using multimedia capabilities in any genre of mass media. At the same time, the objectives of the academic discipline remain a priority, and media educational tasks only contribute to its achievement. Media education technologies assume a multilevel implementation approach. At each stage, an effective mechanism is a popular science film that helps to present complex mathematical material in a fascinating and accessible way. The videos created by students as part of the implementation of the model of media educational development have found their application in classes on the methodology of teaching mathematics, as well as in the implementation of automated solutions. Practice-oriented plots, which have received a visualized format, establish the wide applicability of mathematical knowledge to modern technologies and multiple aspects of everyday life. They serve as an inexhaustible source of actualization and motivation of educational, research and creative potential of students. Having mastered such technologies, future mathematics teachers are able to easily

S. Dvoryatkina (✉) · A. Dyakina · T. Safronova
Bunin Yelets State University, 28, Kommunarov St, Yelets 399770, Russian Federation
e-mail: prorektor-nr@elsu.ru

A. L. Semenov et al. (eds.), *Structural and Technological Transformation of Education in the Post-Pandemic Period*, Lecture Notes in Networks and Systems 1282,
https://doi.org/10.1007/978-3-031-84039-5_5

adapt to the conditions of digitalized education, in which they will have to carry out their professional activities.

Keywords Integrated media education · Media texts · Teaching mathematics · Intelligent learning systems · Motivation of learning

1 Introduction

Modern man lives and develops in a media digital environment, scientists call him homo mediatus [1, 2] and say that for an adequate existence in the digital space, he must become critically and creatively thinking. E.L. Vartanova rightly notes: "A media man is a member of a society whose existence is determined and even shaped by the content produced in the process of mass social communication. The peculiarity of the current period is that a media person is not only a passive consumer of ready–made media products, but also an active participant in the processes of their distribution and even production" [3]. In the field of journalism, the term "participatory culture" is increasingly being used, which defines a different level of communication between the producer and the consumer of information than before. The new digital reality redistributes roles, relies on co-creation with the audience, on its direct participation in the creation of a media product.

Young people function more intensively than other age categories in the media space. However, with all the technical advancement, for the most part she is not ready to critically perceive and evaluate the content of communication flows, is not able to put barriers in the way of negative media influence and protect herself from "information garbage". At the same time, this problem is most often not realized by the youth audience and does not reduce its need for media creation in the public communication sphere.

The youth, especially students, need to learn how to fully live in a saturated digital environment, independently "organize" their own information space. The need to extract from it the necessary professional, socially and personally significant information, satisfy communicative requests, but at the same time protect themselves from malicious and false data. Therefore, the formation of a media-literate younger generation should be part of the education process, and new media–educational approaches that take into account the challenges and demands of the time should be introduced into the educational system itself (at all levels), which is now subject to the process of digitalization.

The present study is devoted to the search for possible ways and methods of integrating media education technology into academic disciplines that have extensive cultural, pedagogical and developmental potential necessary for the formation of a future teacher's media competence.

The purpose of the study is to develop and implement technologies of method education in the process of training future teachers of mathematics for the purpose of creating media texts as effective didactic means of forming stable motives for searching and mastering complex mathematical knowledge.

2 Research Overview

The problems of media education are not new to the Russian scientific community. At different times they were considered by O.A. Baranov, E.A. Bondarenko, L.S. Zaznobina, I.S. Levshina, E.V. Muryukina, S.N. Penzin, A.V. Spichkin, Yu.N. Usov, A.V. Fedorov, I.V. Chelysheva, A.V. Sharikov and many others.

A.V. Fedorov, in particular, defines media education as "the process of personal development with the help and on the material of the media in order to form a culture of communication with the media, creative, communicative abilities, critical thinking, skills of full perception, interpretation, analysis and evaluation of media texts, teaching various forms of self-expression" [4]. Of fundamental importance in this context is the basic concept of "media", which Russian researchers characterize as "an open set of means of artificial and natural, verbal and nonverbal, mass and non-mass communication, which are actualized in the process of communication and significantly affect the quality of information received with their help" [5].

The ideas of integrating media education into academic subjects are being developed in numerous scientific papers [4, 6–10]. However, these ideas have not yet been applied to the process of training future mathematics teachers. At the same time, it is obvious that a modern teacher teaching teenagers should know well media educational technologies, be able to integrate media education into the taught subject and, accordingly, be ready to use the information space in professional activities: to conduct a purposeful search, critical analysis, selection, transformation of media texts, possess the skills to create them. The indicated requirements, dictated by the objective process of digitalization of society, the emerging new way of life ("digital socialization"), the comprehensive process of convergence, can be qualified as competencies that testify to the professional identity of a modern teacher. Therefore, already at the stage of specialist training, "it is necessary to make experimenters out of students, people who are flexible and open to innovation, quickly oriented, quickly adapting. People who will avoid formulaic thinking".

The concept of "*integrated media education*" *should be considered as cultural and pedagogical, directly related to the creative activity of the teacher, with his flexible, non-standard thinking, aimed at the qualitative processing of intellectual abilities, socially significant and cultural qualities of students.*

At the same time, strategically verified technologies for integrating media education into the disciplines and sections taught are of paramount importance. The main priority in this case is the processality of training with an emphasis on the development of critical thinking of the individual, her creative and communicative abilities. The result of such a progressive movement should be the necessary level of media literacy of students, allowing them to realize their professional qualities in the digital economy.

The specific procedural features of the integration of media education into individual disciplines can be formulated as follows:

- The emergence of a new relevant teaching tool—a media text—message containing educational information and having a multimedia character, presented in any form and genre of mass media. The main function of the media text is the transfer of information between the subjects of the educational process–the teacher and the student. At the same time, media (for example, television, video, cinema, the Internet) are used as technical sources for obtaining educational information contained in the media text.
- Priority of the objectives of the academic discipline over media educational tasks. The introduction of a new additional teaching tool in the form of a mass media product should, first of all, contribute to the further development of the discipline taught. At the same time, the following technological procedures are of particular importance:
    a. Finding the "points of contact" of the content of the discipline and the content of external (extracurricular) information, ensuring their overlap;
    b. Formulation of media educational tasks that can be solved within the framework of the selected content of education.
- The manifestation of integration in the use of media texts simultaneously as a means of teaching and as an object of study—information transformed into a new type of knowledge, which causes an increase in the quality of education (thoroughness and awareness of knowledge, developed critical thinking).

The key content features of media education integrated into the process of teaching mathematics are revealed in the following theses:

1. The factual content of the material of the academic discipline of mathematics is at the same time the factual content of media education. For the correct interpretation of information, students need comprehensive knowledge. In this case, it is important to demonstrate the multidisciplinary and supra-disciplinary properties of media education.
2. In the course of educational activities having a media-educational aspect, the basic concepts concerning such a branch of knowledge as informology are mastered by students at the level of intuition and applied in practice.
3. The content of media education is embedded in a specific section of the content of mathematics on the basis of various relationships determined by the following principles:
    - Improvement of learning outcomes;
    - Expediency and possibility of embedding;
    - Availability of methodological tools (techniques, means and forms of media education).

4 The content of media education integrated into mathematics as a subject of teaching presupposes the formation of students' specific skills that meet the

learning goals (search for information; interpretation of the obtained search results and their critical analysis; creation of information transformed into a new type of knowledge), which simultaneously act as the tasks of media education itself.

5 In the content of media education, it is proposed to focus on the activity component, which includes the interactivity of the educational process in the context of the search, analysis and creation of the necessary media product by students. In technological terms, media texts will become the means of integrating media education into mathematics. The construction of integrated media education teaching tools is the understanding and mastering of a certain system of technological procedures. Exactly:

   a. Taking into account the restrictions imposed by the choice of a specific media;
   b. Determination of didactic properties inherent in specific types of teaching aids;
   c. Taking into account the technical capabilities of the selected media;
   d. Determination of the didactic functions of the learning tool (a prerequisite for the procedure: taking into account the correspondence of the didactic functions and technical capabilities of the selected media);
   e. Didactic processing of information for its inclusion in the media text:

o Setting goals for the use of future learning tools;
o Selection of the content embedded in the mathematics program in accordance with the goal setting;
o Selection of additional content in accordance with the goal setting and taking into account didactic principles (relevance, scientific, visibility, accessibility, systematicity).

Media education technologies involve the organization of multi–level activities, including several stages, the first of which is preparatory and motivational [10]. To form stable motives for mastering new things in the cognitive mathematical activity of students, to activate the motivational process for studying complex mathematical knowledge, to illustrate the accessibility and significance of mathematical methods and patterns, mechanisms for implementing intra-subject and inter-subject connections, to implement the student's attitude to self-determination and self–organization– the main tasks of this stage. An effective mechanism in this regard can be a popular science film that helps to present complex mathematical material in a fascinating and accessible way.

3 Results and Discussion

In 2020, the I.A. Bunin Yelets State University (Yelets, Russia) developed a model of media educational development of students, which includes the following components: the information and communication environment of the university, vocational training with integrated media education technologies, additional vocational training in the media education profile [11]. Future teachers of mathematics, along with others, took courses of additional professional training and received the necessary knowledge and skills in the field of media provided by the program. At the same time, the teachers of the courses took into account the profile of the students' future activities, and therefore it was important not only to introduce students to the media sphere from the point of view of "smart use", but also to form a stable need for the transfer of acquired knowledge to students, the desire to work together with students to create media products that can optimize the educational process, lead it is in accordance with modern requests for obtaining complex information through its maximum visualization.

The result of the media education training of students was a collection of videos that illustrated the practical significance of one of the most difficult sections of school mathematics—"Probability and statistics": "Probabilities of probabilities in 3 parts. Part 1. Binomial distribution. Part 2. Bayes' theorem. Part 3. Zero probability does not mean "impossible"; "How a simple formula can simplify life (Bayeux formula)"; "The paradox of birthdays"; "Various applications of normal distribution"; "Probability theory–the dispatcher's tool at the airport" (Fig. 1).

This format of accompanying mathematical information turned out to be the most popular in the educational process, since, on the one hand, it was quite consistent with traditional popular science films widely presented on TV channels in Russia and abroad.

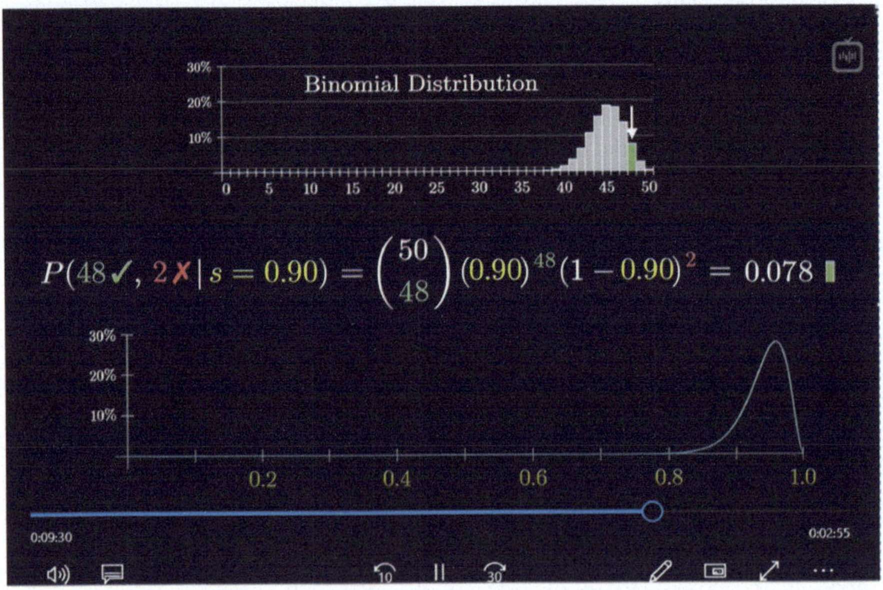

Fig. 1 Film fragment «Probabilities of probabilities in 3 parts. Part 1. Binomial distribution»

The created media products have found their direct application both in classes on the methodology of teaching mathematics, and in the implementation of automated solutions (in hybrid online/offline learning formats, blended learning, in working with fully automated mentor systems).

The inclusion of popular science in the process of teaching mathematics is possible in its various organizational forms, at any stage of the lesson (motivation for educational activities, updating knowledge, understanding the problem and finding ways to solve it, discovering new knowledge, generalizing and systematizing knowledge). Thus, a deeper and more multifunctional mastering of essentially all sections of mathematics, enriched with new qualitative content, characteristics and forms, as well as the individualization of forms and means of pedagogical support in intellectual learning environments, becomes achievable.

In this regard:

- Integration of the objectives of the training course, its content and media education leads to changes within organizational forms, making them flexible and "mobile";
- The use of new teaching tools, the definition of the goals of their application, internal changes in the forms of organization of the educational process, contribute to the change of traditional teaching methods by enriching their internal content with creative methodological techniques that provide specific interpenetration of various methods.

The approbation of the developed media products also allowed us to conclude that the use of integrated media education technology in practical classes in mathematics not only does not contradict, but also presupposes the possibility of using relevant pedagogical technologies, such as project activity technology, research activity technology, problem-based learning technology, game technologies, etc., thereby making more diverse and creative forms of work with students, which in turn allows you to develop interest in the subject and, as a result, motivational sphere of personality.

The integration of media education into the process of teaching mathematics involves an innovative approach of the teacher not only to the collection and processing of educational and extracurricular information, but also to the transfer of knowledge to students. Using educational resources presented in electronic and digital form, as well as technological methods of media education, a mathematics teacher transforms the traditional form of an educational lesson, designs a lesson in an interactive mode, and then implements the project. The future school teacher should be prepared for this kind of activity at the stage of university education.

Visualization of practice-oriented plots of mathematical character establishes a wide applicability of probabilistic and statistical methods both to modern technologies in various fields of knowledge (economics, medicine, psychology, space research) and to multiple aspects of everyday life, and therefore is an inexhaustible source of motivation and actualization of educational-research potential of students.

The second aspect of the introduction of the created media products into the process of teaching mathematics is realized in a multilevel model of the subject area of the hybrid intellectual environment of the support of the research activity of the trainees through the allocation of didactic units (modules) – structured, semantically complete fragments of mathematical content, represented by several levels of meaningful interpretation, differing in the degree of detail and the form of presentation of the material. In the study [12], the main technological stages of the development of schoolchildren's activities in teaching mathematics in a hybrid intellectual system were identified on the basis of the deployment of funding processes in the processes of building the didactic field of educational elements (motivational-value, tentative-preparatory, procedural-activity, evaluative-correctional, generalizing-transformative). The main goal of the motivational-value stage is the formation of positive motivation by creating situations that reveal the historical and genetic aspect of complex knowledge, the new concept being introduced, its interdisciplinary nature, and further practical applicability (Fig. 2).

Inclusion at this stage in the repository of popular science films will demonstrate at an intuitive and visual level a "new" interpretation of a complex mathematical concept, visual modeling of motivationally applied situations, forming a single motivationally applied integrity of complexes of research tasks, in the process of which there is an active mastery of knowledge, a deeper insight into the essence of mathematical concepts, methods, ideas.

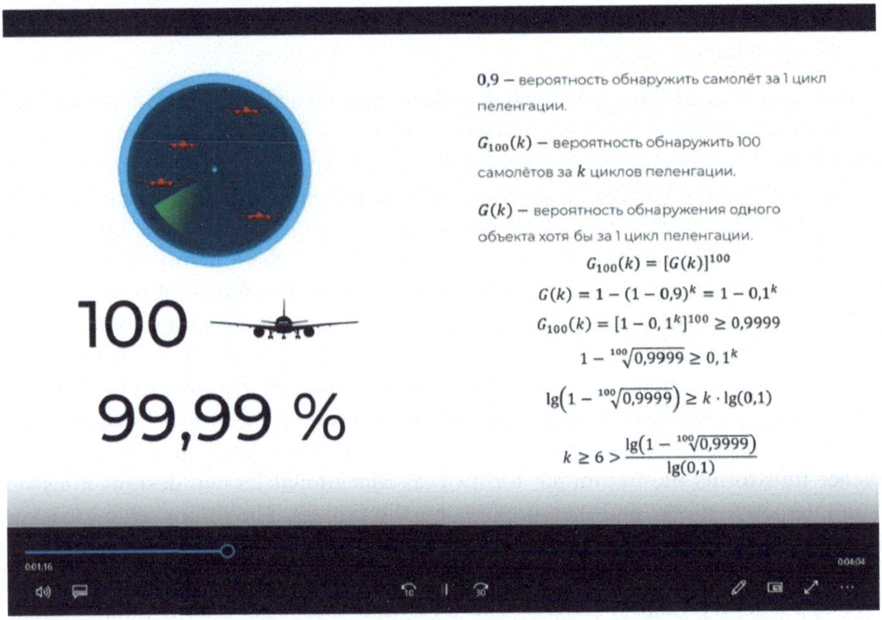

Fig. 2 Bernoulli scheme for radar aircraft detection

4 Conclusion

At the present stage of modernization of the education system in Russia, media education is seen as an integral part of the development of students and an obligatory component of general, secondary and higher education, the need to include it in the content and process-activity structure of mathematical education is due to its wide potential opportunities.

Integration of media education into mathematical disciplines is possible in any organizational forms of the educational process at its various stages, as well as in the implementation of automated solutions.

In our opinion, an important component providing higher professional training of a mathematics teacher is the media competence of a future teacher.

The process of forming the media competence of school teachers of mathematics can be effective as a step-by–step movement from the simple to the complex: from accuracy and meaningfulness of information consumption to analytical fact–checking, then to the development of sustainable media competencies and the formation of critical thinking, and finally to media communication.

Acknowledgements The reported study was funded by RFBR, project number 19-29-14009.

References

1. Mansurova VD (2017) Media Man as a projection of digital ontology: homo mediatus: digital twin or free personality? Palmarium Academic Publishing
2. Pirner ML, Rath M (2003) Homo Medialis. Perspektiven und Probleme einer Anthropologie der Medien, München
3. Vartanova EL (2010) Media education as a priority of social development. Media education at school. MediaMir, Moscow
4. Fedorov A, Levitskaya A (2021) Media manipulation and movies. Media Educ (Mediaobrazovanie) 17(1):64–69
5. Zhilavskaya IV (2021) Systemic and institutional transformations of Russian media education: monograph. MSGU, Moscow
6. Avdonina N, Zaitseva L, Kuprikov N, Kuprikov M (2022) Analysis of media literacy among students of the journalism training program. Media Educ (Mediaobrazovanie) 18(3):327–335
7. Kachkaeva A, Kolchina A, Shomova S, Yarovaya E (2020) Trust, but verify: problems of formation of media literacy and critical thinking of Russian students. Media Pract Educ 21(3):200–211
8. Kazakov A (2021) The world of media literacy education in the focus of alexander Fedorov's scientific school. Media Educ (Mediaobrazovanie) 17(1):77–85
9. McKenzie T (2021) Empowered learning: teaching the video essay in the secondary school environment. Media Pract Educ 22(1):5–6
10. Dvoryatkina S, Dyakina A, Shcherbatykh S (2019) Project technologies in a university's media educational strategy in developing a probabilistic style of thinking for future mathematics teachers. Amaz Investig 8(23):526–537
11. Artemova YuV, Dyakina AA (2021) The strategy of media education development of the future journalist as a factor of professionalism in the digital economy. Quest Theory Pract Jlsm 10(3):543–557
12. Dvoryatkina S, Zhuk LV, Khizhnyak A, Smirnov E, Shcherbatykh S (2021) Open innovation model of student's research activities. J Teach Educ Sustain 23(2):77–90

Geometric Abstractions in Real Space New Opportunities for a Distance Learning and Teaching Geometry

Maria Shabanova, Olga Bezumova, and Svetlana Kotova

Abstract The situation of the Covid-19 pandemic has led to increasing problems in the organization of the educational process. In teaching mathematics, one of the main tasks before the teacher is to teach students to apply their knowledge of geometry to solve actual practical problems. However, the online training mode makes it difficult to complete this task. Online communication makes it difficult for the teacher to deploy real objects that correspond to geometric abstractions. This article presents the technology for teaching geometry, which can be implemented in full-time and distance learning. The article aims to present the technology for teaching geometry and evidence of its effectiveness. We found that learning problems can be solved by using the opportunities provided by new technologies: dynamic geometric environments, augmented reality (AR), and 3D printing. GeoGebra was used as a tool to solve the aforementioned problems because it is a free, fast-paced dynamic geometry system. In 2020 GeoGebra developers provided users with the opportunity to embed interactive dynamic drawings in educational materials using an online platform for distance learning with a feedback option. We have organized experimental geometry training with GeoGebra in schools in the Arkhangelsk region since 2010. In 2020, a new objective arose due to pandemic limitations—the implementation of a remote format system that we had developed. This study shows that the presented technology for teaching geometry allows for strengthening its practical orientation and improves the quality of subject training.

Keywords Secondary geometric education · Dynamical geometry software · GeoGebra · Augmented reality · 3D printing · Distance learning

M. Shabanova (✉) · O. Bezumova · S. Kotova
Northern (Arctic) Federal University Named After M.V. Lomonosov, 17, Severnaya Dvina Embankment, 163002 Arkhangelsk, Russia
e-mail: shabanova.maria-pomorsu@yandex.ru

O. Bezumova
e-mail: o.bezumova@narfu.ru

S. Kotova
e-mail: s.kotova@narfu.ru

A. L. Semenov et al. (eds.), *Structural and Technological Transformation of Education in the Post-Pandemic Period*, Lecture Notes in Networks and Systems 1282,
https://doi.org/10.1007/978-3-031-84039-5_6

1 Introduction

Geometry is extremely important for the intellectual development and formation of practical skills of students. This is emphasized in the papers by [1, 2]. The authors of these works talk about the importance of studying geometry for the development of spatial and logical thinking of students, the formation of their constructive skills and for the formation of mathematical literacy of students. They also note the great difficulties that students face when studying geometry, especially stereometry.

In Russia, geometry has been the subject of study in secondary schools since 1701, when the schools of mathematical and navigation sciences were opened by the decree of Peter the Great. Initially, Euclid's *Elements* was used as a textbook on geometry.

The authors of school textbooks of geometry for a long time followed the traditions of Euclid. The peak of commitment to the scientific rigor and abstractness of the presentation of geometry fell on the Kolmogorov reform of mathematical education [3]. Modern geometry textbooks pay a lot of attention to the issues of visual representation of geometric abstractions and use photographs to demonstrate the connection between geometric abstractions and real objects [4]. However, the results of comparative international studies of students' mathematical literacy also show that the field of "Space and Shape" remains the most problematic for Russian students [5].

2 Survey Methodology

Information technologies are currently opening up new opportunities in solving problems related to learning geometry. The attention of most researchers today is focused on systems of dynamic geometry [6, 7]. Teaching geometry using dynamic geometry systems makes it possible to overcome the lack of spatial thinking development in students, allows implement of the inquiry-based learning, and strengthens the role of experimental methods in teaching geometry. In these articles, it is proposed to use various dynamic geometry systems to teach geometry: Cabri, Geometer's Sketchpad, Cinderella. But most often the authors of the articles suggest using a GeoGebra to teach geometry.

GeoGebra is an open-source software, published in 2002, and it is constantly improving. At first, it only allowed creating the models of planimetric objects. In 2014, the users of GeoGebra became able to create 3D models of geometric objects. In 2017, it became possible to use GeoGebra to create augmented reality and save models in a 3D-printable format. In 2020, the developers presented GeoGebra Classroom, a platform for distance learning enabling feedback while the students work with dynamic worksheets.

The purpose of our research was to develop a technology for teaching geometry using GeoGebra based on the following principles important for the education system in Russia: maintaining a high level of rigor in the presentation of geometric theory in the school geometry course; reliance on a system-activity approach to learning; no additional time spent on learning the software.

We consider the process of mastering each topic of geometry as a gradual mastering of geometric activity in the course of purposeful controlled independent work of students to solve a series of educational and research tasks: (1) on visualization of intuitive geometric images generated by a real object through the construction of a geometric model using computer tools; (2) on studying a variety of real objects described by the constructed geometric model through its dynamization; (3) on obtaining a system of scientific knowledge about the properties of the resulting geometric model by conducting computer experiments with it; (4) on applying the acquired knowledge to the design of new real objects.

We did not make changes to the curriculum and did not write a new textbook. A basic textbook was used to teach geometry. To assess the effectiveness of training with GeoGebra, we used multi-level thematic testing and adapted psychological tests published in a relevant manual. Psychological tests were used to check the structure of motivation to study geometry, the level of educational anxiety associated with geometry, and the level of creativity. The multi-level thematic tests included four tasks:

- Elementary literacy check (skills in recognizing geometric shapes and simple constructions).
- Basic functional literacy check (skills of applying learned solutions in familiar situations without justifying the actions).
- Advanced functional literacy check (ability to justify the possibility of applying the studied techniques in new situations and combining them).
- Creative functional literacy check (ability to create new things based on the acquired knowledge).

For each task, the students were allowed to use GeoGebra or augmented reality technology if they considered it necessary. The students had 45 min to solve problems 1–3. For solving the first problem, students could get 0–20 points, 0–30 points for the second problem, and 0–50 points for the third problem. Solving the 4th problem could bring them additional 0–100 points. Students were allowed to solve it at home.

The teachers and the students did not receive ready-made dynamic models, they had to create them themselves, just as they created static drawings. The students were introduced to GeoGebra tools step by step, as long as they studied geometric shapes. While studying the "Initial geometry information" module, the students mastered such tools as Point, Line, Ray, Segment, Point on Object, Distance or Length, Angle, Angle with Given Size, and Perpendicular Line. While studying the "Triangles" module, they got acquainted with Polygon, Rigid Polygon, Angle Bisector, Midpoint of Center, and Relation. Students learned how to use definitions and theorems to build dynamic models of more complex shapes: adjacent and vertical angles, medians, altitudes, bisectors of a triangle, etc. They had to master the principles of compass and ruler construction to create even more complex models and new tools. For students to master the GeoGebra tools, we have included additional tasks in the practical material of the geometry course.

During the experimental training, we took special steps to avoid the occurrence of the so-called experimental-theoretical gap. The experimental-theoretical gap is a negative effect of studying geometry using dynamic mathematical systems, which has been reported by many researchers [8]. We have developed a special technology of teaching proof with the help of a computer to prevent this phenomenon. It is now applied in teaching geometry in grades 8–9.

This technology involves taking students through three stages in teaching proof:

- Training to conduct confirmatory experiments prior to agreeing with conclusions that are based on exploration experiments.
- Training on logical control of the correctness of algorithms for constructing dynamic models that are used for experiments and constructive proofs.
- Training on the use of a computer as an auxiliary tool for conducting deductive proofs.

We suggested for students to use the special techniques for confirmatory experiments: falling out of the acceptable limits, improving accuracy, reducing the step (Δ).

The course of stereometry for grades 10–11 provided us with ample opportunities to explore the role and ways of using augmented reality technology in teaching geometry. We consider the process of mastering each topic of stereometry course as a gradual mastering of geometric activity through purposeful and controlled independent work of students to accomplish a series of educational and research tasks:

- Visualization of intuitive geometric images generated by a real object through the construction of a geometric model using computer tools.
- Studying the variety of real objects described by the constructed geometric model through its dynamization.
- Obtaining a system of scientific knowledge about the properties of the resulting geometric model by conducting computer experiments with the model.
- Applying the acquired knowledge to the design of new real objects.

We organized the completion of each of the listed educational and cognitive tasks in the form of a game focused on the design and use of the AR model. Here are the examples of some geometric games with augmented reality.

1. "Geometric AR modeling from nature." The participants need to select such architectural landmarks, museum exhibits, street installations or surrounding things that most clearly demonstrate the geometric shape under the study. Students create their own AR model of the object, like artists who draw from nature. The students study its proportions by taking practical measurements. The winner of the game is the student who creates the model that best matches the object in augmented reality. The algorithm of its creation is legitimized and fixed in the constructive definition of the geometric shape.
2. "Animation of geometric abstractions." For its implementation the teacher must select some real objects representing various special cases of the geometric shape under the study. The task is to find objects with the same geometric shape at each

point on the map or terrain plan and to prove the validity of their conclusion by taking a screenshot of the dynamic AR model superposed with the object.

3. "Embodiment of geometric fantasies." During the game, students must use the acquired knowledge to change the reality. This is the game where students play the roles of a scriptwriter, engineer, and cameraman. The game begins with writing a script, where the main character is a real object. Students then design an AR model of the object that allows for predefined actions. They use a model to implement their idea and make a movie.

The game "Embodiment of geometric fantasies" can also consist in preparing a 3D model of a geometric abstraction for 3D printing.

Games with augmented reality helped students better understand the connection of stereometric abstractions with real objects, and see the world through the eyes of a mathematician. In 2020, we had to think about adapting our technology to the conditions of distance and hybrid learning. To arrange the learning process, we used GeoGebra Classroom, Zoom video conferencing, and iDroo, an online whiteboard. The video conferencing system was used as an instrument for communication between teachers and students in the classroom. The online whiteboard allowed the teacher and students to take notes during the lesson. Tasks related to the use of dynamic drawings were placed in GeoGebra Books and then implemented in GeoGebra Classroom. The GeoGebra Classroom allowed teachers to see in real time how students work with dynamic resources.

3 Results and Discussion

The experimental training was designed for five years from grades 7–11 of secondary school. The participants were 704 students of 7–9 graduates and 42 teachers from 29 schools. During the three years of studying planimetry, the students wrote 15 control tests. The results can be seen in Fig. 1.

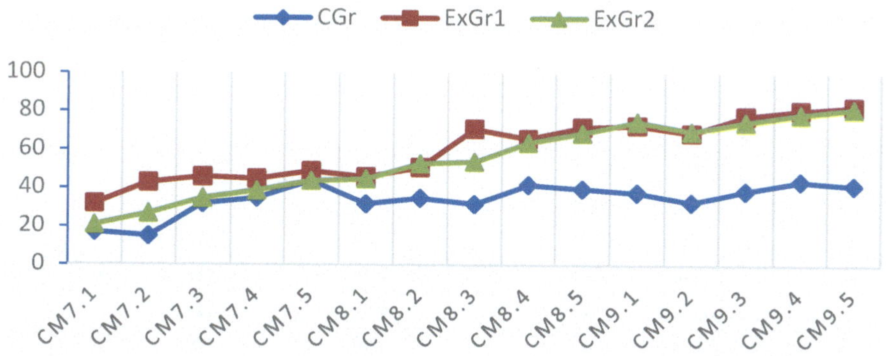

Fig. 1 Results of control measurements

The control group (CGr) consisted of students who studied geometry in the traditional way, that is, without using our technology. Throughout the training, the average result of the group remained at the level of elementary literacy (16–45 points). Experimental group 1 (ExGr1) included those students who not only studied using GeoGebra, but also used this software to perform control test tasks. Experimental group 2 (ExGr2) included those students who were trained using GeoGebra, but did not use this software to perform control test tasks. Figure 10 shows that GeoGebra training allowed students of both experimental groups to achieve higher results. On average, students have reached the advanced of functional literacy (46–85 points), which is characterized by the conscious use of theoretical information in solving problems. Interestingly, starting from grade 9, the results of experimental groups 1 and 2 did not differ statistically from each other.

To analyze the consistency of the distributions, the results of the students were grouped. For each control measurement, the frequency of results related to four controlled levels of functional literacy was determined: elementary (1–15 points), basic (16–45 points), advanced (46–85 points) and creative (86–100 points).

We used the Kolmogorov–Smirnov Test for the hypotheses H0—«The empirical distributions of the results of the control measurement of functional literacy for the two samples (CGr + ExGr1, CGr + ExGr2, ExGr1 + ExGr2) do not differ». The results are presented for the level of significance of 0.05 in Table 1. The critical value of the criterion at this level is 1.36.

The results presented in Table 1 confirm the statistical significance of the differences in the results of CM 7.1 in each pair of samples. This suggests that significant factors at the first stage of the experimental education for the successful solution of the control tasks were training planimetry with GeoGebra, as well as the ability to use this software during the control measurement. At the end of the experimental study period, students who studied planimetry using GeoGebra were more successful than those who studied in the traditional way. The ability to use GeoGebra in solving control tasks was no longer a decisive factor in ensuring the success of students.

Table 1 Results of statistical hypothesis testing

CM	Samples	The maximum of the absolute value of the difference of the accumulated frequency	The value of the criterion	Accepted hypothesis
CM7.1	CGr + ExGr1	0.44	4.53	H1
	CGr + ExGr2	0.29	2.75	H1
	ExGr1 + ExGr2	0.29	3.34	H1
CM9.5	CGr + ExGr1	0.58	5.99	H1
	CGr + ExGr2	0.50	4.69	H1
	ExGr1 + ExGr2	0.08	0.92	H0

Additionally, it would be noteworthy that using GeoGebra ensures 100% academic performance of students in geometry, as well as a gradual shift of their average results towards an advanced level of functional literacy. By the end of the 7th grade, the students were already fluent in GeoGebra. They independently engaged GeoGebra to verify the truth of geometric statements and put forward their own hypotheses based on experiments with dynamical models. We observed a sharp increase in student interest in the study of geometry. At first, it was the result of transferring their interest in computers to geometry, and then it transformed into an interest in making their own geometric "discoveries". The students began to perceive proof as a logical explanation of the observed dependencies.

The technology of stereometry training using augmented reality GeoGebra was tested by us on a small sample. Two 11th grade classes from the pilot schools were selected for the experimental training. There were 24 people in each class. The technology of teaching stereometry using augmented reality was used in the study of two topics in the 11th grade curriculum: "Polyhedrons" (21 h) and "Solid of revolution" (19 h). In Class A, we used our technology when studying the topic "Polyhedrons", and in Class B when studying the topic "Solid of revolution". At the end of the study of each topic, students of both classes were asked to perform the same control test. We also used multilevel tests to evaluate the effectiveness of the technology. An example of this test is presented in [9]. For each task, students in both groups were allowed to use GeoGebra and augmented reality technology if they considered it necessary.

Figure 2 shows that group A exhibited a shift towards higher levels in the first assessment test, while group B had a shift in the second test. These assessment tests were on topics studied using augmented reality technology. Moreover, the results of the second test of group A were slightly worse than the results of the first test of group B. This can be explained by the fact that group A students felt uncomfortable due to the lack of new technologies. These two effects support our assumption about the effectiveness of our proposed augmented reality games in teaching stereometry. Monitoring the progress of students' work showed that computer experiments with 3D models were used by the students with good constructive skills but with a lack of theoretical knowledge.

4 Conclusion

Our goal was to develop a technology for teaching geometry using GeoGebra, which is consistent with the traditional deductive construction of a school geometry course in Russia, dividing it into planimetry and stereometry, and at the same time solves some student's problems associated with the learning of this course: prolonged isolation of student's knowledge in geometry from the areas of their practical application; inability of students to create and operate planimetric and spatial geometric models; the problem of the experimental-theoretical gap. The technology is based on the fouling ideas: inquiry-based learning, mastering the tools of the GeoGebra program

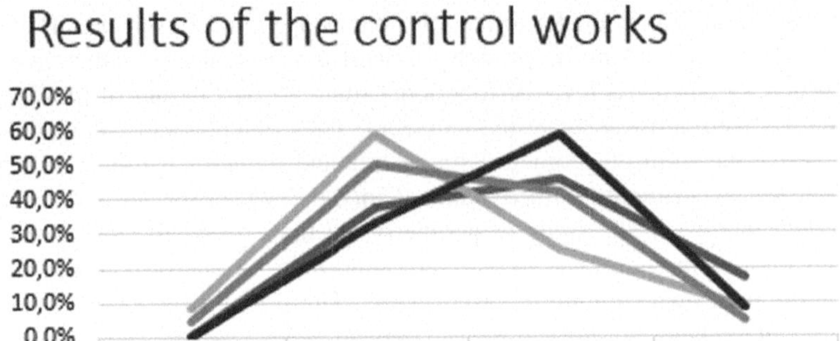

	Elementary level	Basic level	Advanced level	Creative level
Group A CW1	0,0%	37,5%	45,8%	16,7%
Group A CW2	8,3%	58,3%	25,0%	8,3%
Group B CW1	4,2%	50,0%	41,7%	4,2%
Group B CW2	0,0%	33,3%	58,3%	8,3%

Fig. 2 Results of the control works students of the 11th grade

in the context of studying a school geometry course, giving students the opportunity to use the GeoGebra as a computer environment of learning activity, strengthening the role of constructive and experimental methods in the study of geometry.

Our research has shown that the inclusion of students in the creation and use of dynamic interactive models of geometric abstractions and real objects increases their educational results. We believe that the proposed technology allows us to avoid the appearance of a theoretical–experimental gap. This is confirmed by our experiment. We regret that so far not all students of Russian schools can use augmented reality as an auxiliary means of studying stereometry. This is due to the fact that the application ARGeoGebra is still stable only on operating systems iOS. But we think that this problem will be solved very soon by the international GeoGebra development team.

Acknowledgements We are extremely grateful to the organizers of the project, Professors S. Grozdev, T. Sergeeva, and also to the teachers and students of the experimental schools, who made an invaluable contribution to the implementation and development of the project ideas.

References

1. Develi HM, Orbay K (2003) Why and what kind of geometry teaching in primary education. J Natl Ed 157:115–122
2. Vakaliuk TA, Shevchuk LD, Shevchuk BV (2020) Possibilities of using AR and VR technologies in teaching mathematics to high school students. Univers J Ed Res 8(11B):6280–6288
3. Kolmogorov AN, Semenovich AF, Cherkasov RS (1979) Geometry. Textbook for grades 6–8 of secondary school. Prosveshcheniye, Moscow

4. Atanasyan LS, Butuzov VF, Kardomtsev SB (2010) Geometry grades 7–9. Textbook for general education institutions, 20th ed. Prosveshchenie, Moscow
5. PISA Results, volume I–III (2018). https://www.oecd-ilibrary.org/education/pisa_19963777. Accessed 24 Mar 2022
6. Almeida J, Castro M (2021) Applying geogebra in descriptive geometry online teaching to model fundamental concepts. In: (Blucher Design Proceedings, Org., XXV SIGRADI 2021, São Paulo, Anais, 2021. Editora Blucher, São Paulo) Paper presented at the XXV international conference of the Ibero-American society of digital graphics 1:389–400
7. Hannafin RD, Burruss JD, Little C (2001) Learning with dynamic geometry programs: perspectives of teachers and learners. J Ed Res 94(3):132–144
8. Hanna G (2000) Proof, explanation and exploration: an overview. Ed Stud Math 44:5–23
9. Shabanova M, Bezumova O, Zatsepina E, Malysheva S, Kotova S, Ovchinnikova R (2020) Learning stereometry in a secondary school within GeoGebra's augmented reality app. J Phys: Conf Ser 1691:012115

Methodical Support of Hybrid Intellectual System with Cognitive Modeling of Engineering-Research Activity on the Basis of Thinking Profiles

Irina Karpacheva, Elena Igonina, and Galina Simonovskaya

Abstract Development of methodical support of hybrid intellectual system with cognitive modeling of design and research activity of schoolchildren in mathematics. Construction of this system is carried out on the basis of integration of thinking profiles, modified with the development of learning technologies in the context of digitalization of education, and typology of levels of project-research activities of schoolchildren in intellectual knowledge environment (search and reproductive, empirical, theoretical). When forming the content of the system were taken into account the personalized profiles such as: symbolic, geometric, specific, computational, historical and genetic. The filling of the system was determined by research tasks on the main sections of the topic «Complex numbers». Tasks were selected in the context of cognitive characteristics and principles of personalized profiles, their accuracy and ability to reflect the student's preferences and needs.

Keywords Hybrid intellectual system · Design and research activities · Thinking profiles · Research tasks

1 Introduction

Prospects of solving problems of project-research activity management of students in modern conditions are connected with creation of automated learning systems based on artificial intelligence techniques [1, 2]. Intellectual management of project-research activity of students in mathematical education is designed to provide achieving high quality education through individualization. The effectiveness of intelligent learning systems depends on principles of personalized profiles, their accuracy and ability to reflect user's preferences and needs. Currently developed intelligent systems in which personalization is achieved, are based on cognitive modeling of personal characteristics, affective states, learning dynamics, learning

I. Karpacheva · E. Igonina (✉) · G. Simonovskaya
Bunin Yelets State University, 28, Kommunarov St, Yelets 399770, Russian Federation
e-mail: kmaiem12@mail.ru

A. L. Semenov et al. (eds.), *Structural and Technological Transformation of Education in the Post-Pandemic Period*, Lecture Notes in Networks and Systems 1282,
https://doi.org/10.1007/978-3-031-84039-5_7

styles, creativity. Cognitive modeling allows a priori to predict the results of possible managerial decisions and to choose the most effective solution, to develop an adaptive learning quality management mechanism [3].

The idea of designing intelligent learning systems based on learning styles is recognized by the academic community as dominant. Productivity and performance of adaptive systems, based on learning styles, quality assurance and optimization of learning are noted in a number of studies. One of the most common style-based models is the Felder-Silverman model. This model is based on information acquisition and processing methods and takes into account 4 factors with opposite values: sensory/intuitive (perception mode); visual/verbal (presentation mode); active/reflexive (processing method); sequential /holistic (organization method) [4].

However, the problem of creating a holistic concept of personalized adaptive learning in an electronic environment based on the principles of hybridity, interactivity, intelligence management remains relevant.

The concept of intellectual management of design and research activities of schoolchildren in the hybrid learning environment of mathematics proposed in this paper is based on the cognitive profiles of the learner and typologies of levels of activity of schoolchildren in the intellectual environment of knowledge. Profiles are concretized taking into account the modern tendencies of digitalization of education for students, motivated to study mathematics. Diagnostics of thinking profiles is carried out with the help of the technique «Thinking Profile» (Ganzen et al.) [5].

2 Research Overview

The study of the phenomenon of individual thinking is carried out in psychology since the middle of the twentieth century. Thinking skills, along with learning and complex problem solving based on understanding, refer to the basic skills of the XXI century [6, 7]. Cognitive psychology is currently researching thinking styles, types, and profiles. The style of thinking is defined as the functional organization of the emergence and development of cognitive neologisms, their stable manifestations in humans [8] Type of thinking means individual way of analytic-synthetic transformation of this information [9]. Thinking profile characterizes the dominant ways of intellectual processing of information, which is based on the inherent types of thinking and level of creativity» [5]. In classical psychology such basic types of thinking are defined as: Subject, figurative, iconic and symbolic [10]. Based on D. Bruner's thesis on thinking as a translation from one language to another, researchers expand the set of types of thinking: subject-form (practical), object-symbolic (humanitarian), object-symbolic (operator), figurative (artistic), figurative-symbolic (technical), iconic-symbolic (theoretical).

The thinking profile is the most important personal characteristic of the student. It displays not only the dominant ways of processing information children and the level of their creativity, but also determines the style of activity, inclinations, interests and professional orientation of the person. This paper provides an example of a hybrid intellectual system with cognitive modeling of design and research activities of schoolchildren who are motivated to study mathematics, in this discipline. Filling of the system is carried out on the basis of the following profiles of thinking, modified by E.I. Smirnov [11] taking into account the development of learning technologies in the context of the digitalization of education:

1. *Iconic-symbolic profile.* The dominant type of thinking is symbolic, abstract-theoretical thinking. In the process of solving abstract theoretical problems, the student reveals the regularities of the problem or phenomenon in abstract concepts, resorts to operations performed in the mind (not directly dealing with the experience gained through the senses). Iconic-symbolic profile is typical for students who are engaged in scientific theoretical research.
2. *Figure-geometric profile.* Dominant thinking type is figurative-symbolic. Figure-geometric profile is connected with the learner's ability to understand the drawings, to understand the schemes of technical devices and their work, solve practical physical and technical problems. It is typical for students inclined to choose a profession such as «man-technique».
3. *Concrete activity profile.* The dominant type of thinking is subject-form. The thought process is carried out directly in the perception of the surrounding reality and the practical transformation of real objects in real time. Necessary images are presented in short-term and operative memory. This profile is noted among students engaged in productive work, which results in the creation of a material product.
4. *Information-computing profile.* The dominant type of thinking is object-symbolic. The profile is based on reproductive thinking, aimed at the execution of selected short-term operations according to a given scheme with the use of short-term and operative memory. It is typical for students who like to do counting or computing, accounting work.
5. *Historical-genetic profile.* Dominant types of thinking is object-symbolic and figurative-symbolic [12]. Object-symbolic type occurs in children of humanitarian orientation, figurative-symbolic type is more prevalent among schoolchildren engaged in some kind of art. The profile is connected with the discovery of new knowledge, with the generation of original ideas.

It is worth noting that the profile of thinking is determined not only by the dominant type of thinking, but also by the creativity of the student. Task facets used to fill the knowledge base of a hybrid intellectual system, are needed to be built, including on the basis of cognitive modelling of the design and research activities of schoolchildren. Mathematics tasks should be offered taking into account the following levels of design and research activity of students in the intellectual environment of knowledge: (1) search and reproductive; (2) empirical; (3) theoretical [12].

3 Results and Discussion

Let us illustrate the richness of the hybrid intellectual system taking into account thinking profiles and specified levels on the example of realization of project-research tasks by students on the topic «Complex numbers». The choice of this topic is due to the fact that complex numbers are an interesting subject for study, which provides the key to solving many applied problems that cannot be considered within the narrow limits of elementary mathematics. When solving problems using complex numbers, new interesting facts and generalizations are sometimes found [13].

Logically it would be right to finish the study of the concept of number in the school with the topic «Complex numbers». In the history of domestic mathematical education there were periods, when it was part of the high school math program. Beginning in 1917, Soviet schools studied purely imaginary numbers, since 1932, the topic «Complex numbers» has been included in the program of the 8th grade, the subject continued until the end of the mathematics course in school. However, since 1968, as a result of modernization of the general education course of mathematics, the topic «Complex numbers» in general education classes was not considered, and its study was kept only in major classes. It should be noted that sometimes this topic was also reflected in the school course of mathematics in LH. It is also worth noting that complex numbers are important as a separate field of mathematics, in which the knowledge and skills acquired by students of algebra and trigonometry work perfectly [14].

In this paper, the learning of the following sections has been taken into account to determine the base content of the hybrid intellectual system: problems leading to the concept of complex number, definition and forms of writing complex numbers, operations on complex numbers represented by different forms of notation, geometric interpretation of complex number. For the development of methodical support of hybrid intellectual system the above profiles of thinking were used (by E.I. Smirnov) taking into account their specific characteristics and levels of project-research activity of students in the intellectual environment of knowledge. The number of database tasks is represented by double numbering: the first digit is the profile number, the second is the level of knowledge. Depending on the level, volume and stage of students' learning of the studied material facets of tasks can be correlated.

Profile 1 (iconic-symbolic) can be presented with the following tasks:

Task 1.1. Find the quotient of complex numbers in two ways $\frac{1+3i}{1+i}$: by definition and using complex conjugate number.

In the course of the solution the first way learner using the proportion $\frac{1+3i}{1+i} = a+bi$, goes on to solve the equation system $\begin{cases} a-b=1, \\ a+b=3 \end{cases}$. To find the quotient of complex numbers by the second way the student uses a complex conjugate number of the denominator of the given fraction. The comparison concludes that second method is more rational and requires no special formulas to be memorized.

Task 1.2. Solve the quadratic equation with the unknown z:

$$z^2 + (-5 + 6i)z - (1 + 9i) = 0$$

When solving a square equation, the student first calculates the square root of the discriminant, which is a complex number. Since root extraction from a complex number results in a complex number, then root of discriminant is sought as $D = a + bi$. In our case it is $D = -7 - 24i = a + bi; 7 - 24i = (a + bi)^2$, then $7 - 24i = a^2 + 2abi - b^2$. Using the knowledge that two complex numbers will be equal if their actual and imaginary parts are equal, respectively, the learner must solve the system to find unknown values a and b. Considering any of the pairs received $a_1 = 3, b_1 = -4$ or $a_2 = -3, b_2 = 4$, for example, the first, it turns out that $\sqrt{D} = 3 - 4i$, and then $Z_1 = 4 - 5i$, $Z_1 = 1 - i$.

Task 1.3. Argand Française symbols can be used to write complex numbers in trigonometric form more briefly. These symbols were first used in 1813 by Française, a professor of artillery at Metz. Finally this designation was accepted by Cauchy. The complex number z, which amplitude is r, $r > 0$, and the main value of the amplitude is φ $(-\pi < \varphi \leq \pi)$ is denoted by r_φ, that is $r(\cos\varphi + i\sin\varphi) = r_\varphi$. Complex number with $r = 0$ doesn't have an amplitude. Positive real number X, will be written as X_0, $X > 0$, and minus number will be written as $(-X) = X_\pi$, $X > 0$, unit imaginary number will be written as $i = 1_{\frac{\pi}{2}}$. . Write down the known rules (addition, multiplication, division) of the actions on complex numbers $z_1 = r_\alpha$ and $z_2 = \rho_\beta$.

It is also possible to invite students to present formulas for the expression of sine and cosine through Argand formulas. Solving this task, the student analytically works with complex numbers written in different forms. Symbolic record, its interpretation allows on the one hand to ignore the essence of the tasks performed, but the resulting formulas can then be used to consider trigonometric functions.

Profile 2 (figure-geometric) can be filled with the following tasks.

Task 2.1. Find the loci Re $z > c$, where z is a complex number, c is real number.

When working with this task, the student first determines the real part of the complex number $z = x + iy$. The generalization of the gained knowledge on the topic «Solution of inequalities» is made further and then moving to the structure $x > c$. The resulting inequality is the simplest, however, you need to consider the feature that c can run all real numbers, but the response structure is not affected (see Fig. 1): $X \in (c; +\infty)$.

Task 2.2. Find the loci $|z - z_0| < c$, where z, z_0 are complex numbers, c is real number.

When solving this problem, the student needs to go to the algebraic form of complex numbers: $z = x + iy$, $z = a + ib$. When inequality is transformed, inequality becomes the inequality, which defines the inner region of the circle with the center

Fig. 1 Real number values c

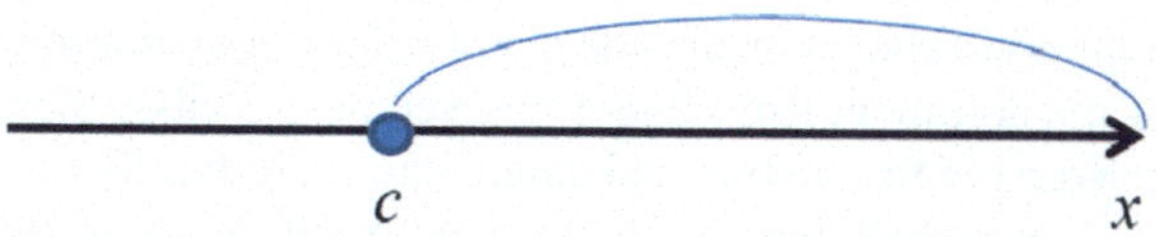

Fig. 2 Loci

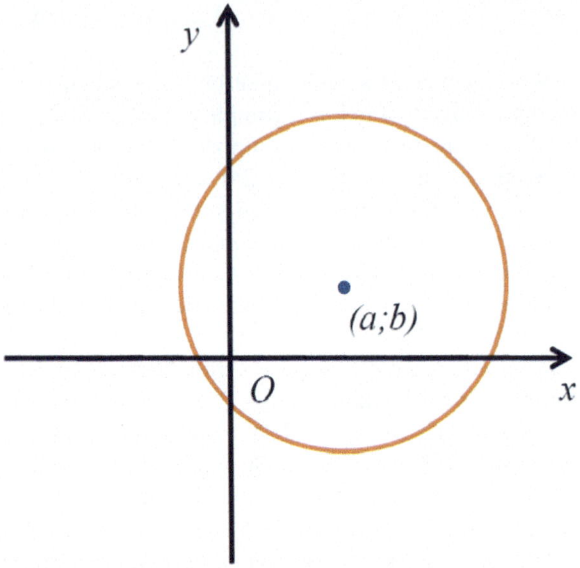

at the point (a; b) and radius c (Fig. 2). At this stage, a study should be carried out: impact of values z_0 and c on the result of solution.

Task 2.3. Consider vector interpretation of addition operations and subtraction of complex numbers.

During the task the student uses the knowledge that each point M, matching the complex number z, on the plane of complex numbers is uniquely matched by the vector $\overline{\text{OM}}$ starting at zero point O. The addition and subtraction of complex numbers written in algebraic form clearly corresponds to the addition and subtraction of corresponding vectors. That is, if a and b are complex coordinates of points A and B, then number $c = a + b$ is the coordinate of point C according to Fig. 3. The complex number d = – b is corresponded by point D.

Students can also be asked to consider the geometric interpretation of the multiplication operation of complex numbers represented in trigonometric form. Geometric imaging tasks are examined with the support of an interactive environment that allows to perform dynamic images.

Profile 3 (*Concrete activity)* can be represented by practical objectives.

Task 3.1. For example, one of the most visible interpretations, geometric is points in the plane, whose coordinates correspond to the actual and imaginary part of the complex number. Considering some curve as a population of such points, one can describe it as a complex function of the real argument.

For example, when considering a function f (t) = f(t) = i^t, by explanating t as temporal value, it results the degree of rotation of unit vector. If you change the value of the argument from 0 to 4, the vector describes the complete circle. It is possible conversion to function and canonical, more familiar:$i^t = e^{\frac{i\pi t}{2}}$.

A graphical dynamic representation of functions of this kind is also possible.

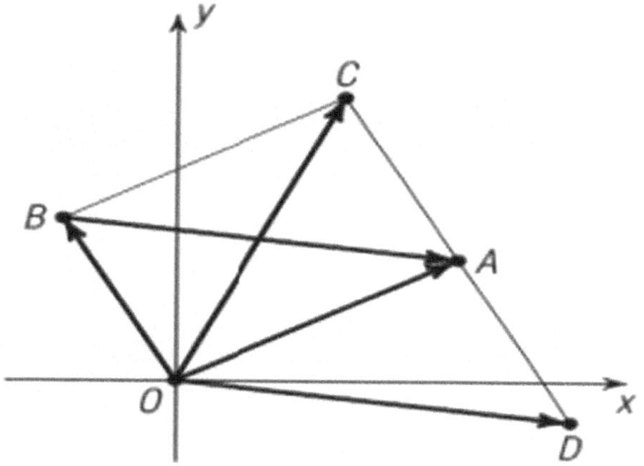

Fig. 3 Vector interpretation of addition and subtraction operations of complex numbers

Another practical use of complex numbers is the use of the numbers in solving planimetric problems. It should be noted that further consideration of linkages will be required.

Task 3.2. The triangle ABC is refined in a circle centered at the origin. The triangle A'B'C' is symmetrical with respect to the actual axis. Prove that their orthocenters (height intersection points) are symmetric with respect to the actual axis.

A solution using complex numbers can be presented as follows.

Let the points of ABC have complex coordinates $A(z_1)$, $B(z_2)$, $C(z_3)$. Then the complex coordinate of its orthocenter is $H(z_1 + z_2 + z_3)$. The triangle A'B'C' will be refined in the same circle (by definition of symmetry and by property of diameter of circle, perpendicular to chord), complex coordinates of its points are respectively $\overline{z_1}$, $\overline{z_2}$, $\overline{z_3}$. Complex coordinate then H' will be $\overline{z_1}+\overline{z_2}+\overline{z_3}$, but for complex numbers the $\overline{z_1} + \overline{z_2} + \overline{z_3} = \overline{z_1 + z_2 + z_3}$. As was to be proved.

In the next level tasks it is advisable to reflect the application of complex numbers, for example, application in the economy (in particular in financial transactions when considering capital formation).

Task 3.3. Let's look at the rise at the complex interest rate given by the formula:

$$FV = PV(1 + r)^t \tag{1}$$

where the initial amount PV is extended over a period of t at the rate of r – complex number of type $a + bi$, FV *is* final amount. Knowing the interest rate $r = a + 1{,}09i$, it is necessary to construct a schedule of growth and determine the period of capital accumulation and the rate of capital growth for this period.

To solve the problem, the student must first consider the trigonometric form of representation of the complex number z, which participates in the formula of accretion (1):

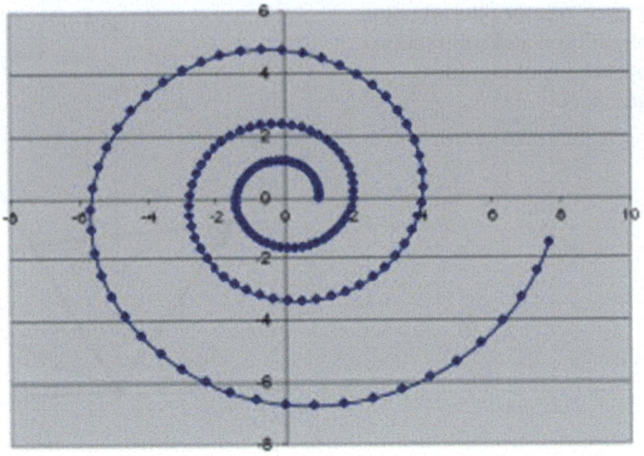

Fig. 4 Complex sum *FV* accumulation schedule at the rate of $r = a + 1{,}09i$

$$z = 1 + r = (1 + a) + bi = mod(\cos\alpha + i\,sin\,\alpha),$$
$$\text{mod} = \left((1 + a)^2 + b^2\right)^{1/2},\ \alpha = arctg\frac{b}{1 + a} \quad (2)$$

According to the Moivre formula, accretion (1) looks like

$$F\,V = PV\ \ \text{mod}^{\,t}(\cos\,t\alpha\ +\ +i\ \sin\,t\alpha)$$

Complex component of number *FV* can be discarded and considered as a real monetary sum only the real part. Note that the schedule of accumulation of the complex sum *FV* in the complex plane is a spiral which intersects with the real axis in its positive and negative part. At compound rate $r = a + 1{,}09i$ according to the formulas (2) we will get a gain graph at the Fig. 4. The capital accumulation period is 5 years, the capital growth rate for this period is equal to 2.

As an alternative task, one can consider the use of complex numbers in physics. For example, in electrical engineering [15] real numbers are used to measure resistance. It is the property of the object to prevent electric current from passing through it and imaginary numbers are used to measure inductance (ratio of magnetic flux to current in coil) and capacitances (ratio of electric charge to potential difference between capacitor plates). Application of complex numbers allows to use laws, formulas and methods of calculation, which are used in DC circuits to calculate AC circuits. It is possible to invite students to consider other applications of complex numbers.

For the Profile 4 *(informational-computing)* could be offered the tasks of such a plan as:

Task 4.1. Using the properties of complex numbers.

$|z_1 + z_2| \le |z_1| + |z_2|$ and $|z_1 + z_2| \ge ||z_1| - |z_2||$,

show, that $7 \le |z + 6 - 8i| \le 13$, if $|z| = 3$.

When working with this task, the student tries to see the application of the proposed properties to accomplish the task in a context. Namely, the first property is implemented as follows:

$|z + 6 - 8i| \leq |z| + |6 - 8i|$, where $|z| + |6 - 8i| = 3 + \sqrt{6^2 + 8^2} = 3 + \sqrt{100} = 13$,

second property application:

$|z + 6 - 8i| \geq ||z| - |6 - 8i||$, where $||z| - |6 - 8i|| = |3 - \sqrt{6^2 + 8^2}| = |3 - \sqrt{100}| = 7$,

given the received we will have:

$7 \leq |z + 6 - 8i| \leq 13$.

Task 4.2. Develop a program (or algorithm) to solve the following problem. Complex numbers are given $a = \alpha + \beta i, b = \gamma + \delta i, c = \sigma + ui$. Find complex number $d = \varphi + \omega i$, represented by formula $d = a^2 * \frac{a+b}{a-bc}$.

Task 4.3. Express cos5φ and sin5φ in terms of cos φ and sin φ, using a combination of the Newton binome formula and the Moivre formula.

The combination of the Newton binome formula and the Moivre formula is inexhaustible source of combinatorial and trigonometric identities.

According to the Newton binome formula we have:

$$(\cos\varphi + i\sin\varphi)^5 = \cos^5\varphi + 5i\cos^4\varphi\sin\varphi +$$
$$+ 10i^2\cos^3\varphi\sin^2\varphi + 10i^3\cos^2\varphi\sin^3\varphi + 5i^4\cos\varphi\sin^4\varphi +$$
$$+ i^5\sin^5\varphi = \cos^5\varphi - 10\cos^3\varphi\sin^2\varphi + 5\cos\varphi\sin^4\varphi +$$
$$(5\cos^4\varphi\sin\varphi - -10\cos^2\varphi\sin^3\varphi + \sin^5\varphi)i$$

According to the Moivre formula at $n = 5$ will have:$(\cos\varphi + i\sin\varphi)^5 = \cos 5\varphi + i\sin 5\varphi$. Therefore we have,$\cos 5\varphi = \cos^5\varphi - 10\cos^3\varphi\sin^2\varphi + 5\cos\varphi\sin^4\varphi$, $\sin\varphi = 5\cos^4\varphi\sin\varphi - 10\cos^2\varphi\sin^3\varphi + \sin^5\varphi$.

To the Profile 5 (historical-genetic) can be attributed tasks in which complex number is «covered» by historical layer. It is necessary to open it and understand the mathematical component, for example:

Task *5.1.* Conduct research on Argand symbols.

During the assignment, students need to learn more than just historical information about the works of the French mathematician Argand (1768–1822), but the history of the introduction of symbols developed by scientists to denote complex numbers. To test knowledge on the studied topic in Task 5.1 you can offer specific test questions.

For the second level, it is possible to use tasks related to the study of complex numbers and their relationship with different types of transformation..

Task 5.2. Let this transformation *T* be operation that each complex number *z,* that is, each point of the plane, aligns the other point *T*(*z*). Let's consider the photo on the plane of any size and show its image, obtained by transformation: each pixel of a photo will pass to the place of the plane, where its transformation carries over. For example, to display *T*: *T*(*z*) = *z*/2 each number is divided by 2. Therefore, the photo will just shrink in half - this is compression. This transformation in geometry is called homothety. To map T: $T(z) = iz$ the photo is rotated counterclockwise by 90° by definition of the number i.

Students can be asked to perform such transformations as:

(1) $T(z) = (1 + i)z$. Since the number module $1 + i$ is equal to $\sqrt{2}$, and the amplitude is 45°, then the transformation is a composition of turning the preimage by 45° and homothetics with coefficient $\sqrt{2}$ – this transformation is called uniformity. The great advantage of complex numbers is that they can describe uniformity transformations simply as multiplications by numbers.

(2) $T(z) = -1/z$. This transformation is close to the inversion, namely to the origin of the coordinates, which corresponds to the number 0, this transformation cannot be applied, but we will agree to assume that point 0 goes for infinity. The motivation is very simple: when a complex number z approaches 0, that is, its module tends to 0, its image module $|-1/z|$ is attached to the module z, which means he's going for infinity. So, this transmutation in zero has a «blast»: the little neighborhood of zero goes very far beyond the screen. Conversely, points that were very far from zero, retract, moving to points very close to zero. The main property of inversion is that it turns circles into circles or lines. Artists often use transformations of this kind, calling them anamorphoses.

The next level of research tasks is devoted to acquaintance with the views of P.A. Florensky on suspicion in geometry. The content of 5.3 tasks can be related to the study of the system of the most common complex numbers (double, dual, ordinary complex numbers, quaternions, octaves, hypercomplex number system). Students can also be asked to continue with the 5.2 task to conduct research work on the topic: «Complex numbers and fractals». The proposed assignments relate to a rather complex profile and imply a high level of learning.

4 Conclusion

The proposed approach to the definition of training profiles allowed to develop a methodically substantiated system of tasks to fill the knowledge base of the hybrid intelligent support system of design and research activities of schoolchildren in mathematics based on cognitive modeling When developing the system of assignments, cognitive abilities of students were taken into account, based on the dominant type of thinking and creativity. The set of tasks for each profile was determined taking into account their implementation by students at three levels of project-research activities in the intellectual environment of knowledge: research, and reproductive, empirical and theoretical. During the course of the work, a knowledge base was developed and presented by five clusters of three levels in each cluster. Informative filling of hybrid intellectual system was carried out on the example of realization of project-research tasks on the topic «Complex numbers», which took into account the mastery of the main sections of the topic. This approach to intellectualization of learning systems based on individual needs and learning styles was proposed by authors for the first time.

Acknowledgements The reported study was funded by RFBR, project number 19-29-14009.

References

1. Khasbulatova B (2016) Principles of building intelligent learning systems. E-learning in continuing education 91:957–962
2. Alyosheva L (2018) Intelligent learning systems. Bulletin of the university 1:149–155
3. Grechko M (2017) Cognitive modeling as a tool of adaptive quality management of education. Natl Interes: Priorities Secur 4(349):1–16
4. Samia D, Badji M (2016) An adaptive e-learning system based on student's learning styles: an empirical study. Int J Distance Educ Technol 14(3)
5. Ganzen V, Malyshev K, Oginets L (2001) Profile of thinking. A workshop on the psychology of professional activity. Peter, St. Petersburg
6. Ertmer P, Newby T (2013) Behaviorism, cognitivism, constructivism: comparing critical features from an instructional design perspective. Res Updat Key Train Mentor Top 26(2):43–71
7. Hall M, Farkas J (2013) Cognitive skills and attitude behavioral characteristics in adolescence and the level of wages. Quest Educ 3:5–57
8. Belousova A (2015) Development of a personal potential in collaborative thinking activity. Procedia St. Petersburg—social and behavioral sciences. In: Proceedings of the 5th ICEEPSY international conference on education and educational psychology 171:987–994
9. Malyshev K (2014) The study of thinking using a multidimensional typological approach. Yaroslavl Pedagogical Bulletin 2(2):218–222
10. Bruner J (2008) Psychology of cognition. Beyond the immediate information: monograph. Direct-Media, Moscow
11. Smirnov E, Dvoryatkina S, Shcherbatykh S (2020) Parameters and structure of neural network databases for assessment of learning outcomes. Int J Criminol Sociol 9:1638–1648
12. Dvoryatkina S, Smirnov E, Shcherbatykh S (2021) Intellectual support of project and research activities of schoolchildren in a hybrid environment of teaching mathematics. Yelets
13. Ouchi M (2019) Entertaining Mathematics. DMK Press, Moscow, Complex Numbers
14. Voitenko T, Tsygankova A (2019) Methodological features of studying the topic complex numbers in specialized classes of high school. Mod Probl Sci Educ 4:7–14
15. Mizinova L, Popova S (2018) Application of complex numbers in electrical engineering. Int Stud Sci Bull 3(1):320–323

Transformation of STEM Education at University Level

Features of Distance Learning Lessons on the Methodology of Teaching Mathematics for Bachelors in the Context of the COVID-19 Pandemic

Marina V. Egupova

Abstract The COVID-19 pandemic necessitated the transfer of students' education to a distance learning format, which at that moment most universities in Russia were not ready for. The experience of organizing the educational process in the current conditions, which allowed us to move to a higher level of digitalization of education, is investigated. The issues of methodological training of 3rd-year undergraduate students, future mathematics teachers are considered. The period of adaptation to the digital format of undergraduate students' education is described. A decrease in the level of cognitive activity and the level of mastering of educational material in the discipline "Methods of teaching mathematics" was stated. The difficulties of a technical, organizational and substantive nature that accompanied the emergency transition to distance learning are identified and described, and methods for overcoming them are proposed. The features of the preparation and conduct of training sessions on the discipline "Methods of teaching mathematics" using the electronic information and educational environment InfoDa (LMS Moodle) of Moscow Pedagogical State University are highlighted, which can be considered in developing other training courses on similar topics.

Keywords Teachers training · Undergraduate student · Distance learning · Methodology of teaching mathematics · COVID-19 pandemic

1 Introduction

The pandemic has made significant changes in the educational process of higher education institutions in Russia. The introduction at the end of March 2020 of forced restrictions due to the increase in the incidence of the new coronavirus infection (COVID-19) and the need to continue the educational process led to the transfer of student education to a distance format. By this time, in Russian universities,

M. V. Egupova (✉)
Moscow State Pedagogical University, 1/1, Malaya Pirogovskaya St, 119435 Moscow, Russia
e-mail: mv.egupova@mpgu.su

A. L. Semenov et al. (eds.), *Structural and Technological Transformation of Education in the Post-Pandemic Period*, Lecture Notes in Networks and Systems 1282,
https://doi.org/10.1007/978-3-031-84039-5_8

this format of education was not widespread, and the overwhelming majority of students and teachers did not have relevant experience [1]. Many teachers had experience of working in a mixed format, in which they used distance learning technologies to control or organize independent work of students. However, the set of educational materials in electronic form did not allow to switch to distance learning simultaneously.

Nevertheless, in the educational systems of many countries, distance learning in universities is widespread and is the subject of numerous scientific studies. The main advantage of such training is its integrability, unhindered access to the information and educational environment of another university, to world information resources, as well as the opportunity to study with leading teachers in the chosen specialty as part of the joint implementation of educational programs by universities [2].

Along with the obvious advantages of distance learning, researchers note difficulties in organizing independent work of students, in motivating them to study, in digital tools choice, as well as lower performance compared to traditional classroom learning.

It is also emphasized that the teacher must have several special competencies to work in the conditions of using digital technologies for educational purposes. In particular, one of the important competencies is the creation of an electronic course in an electronic information and educational environment, namely, the development and filling of such a course with special content that allows, in the absence of face-to-face contact with students, to maintain the required quality of teaching and ensure the achievement of the planned learning outcomes.

The article written by S.D. Karakozov, E. E. Kovalev, V. G. Manyakhina, O. V. Muravyeva, A.V. Nikiforova, K. S. Smotryaeva [1] presents how these problems were solved at the Institute of Mathematics and Informatics of the Moscow State Pedagogical University. The authors, based on a survey of teachers and students, presented an analysis of the process of adaptation to a distance learning format and the choice of technologies for conducting online classes.

Obviously, at the moment, due to limited access to full-time education, there are unique conditions for the emergence and testing of new teaching methods in higher pedagogical schools, which is confirmed by many studies [3]. We will present our own teaching experience of distance learning of third-year undergraduate students in the discipline "Methods of teaching mathematics" in the same period. This discipline is included in the mandatory part of the bachelor's degree program.

Thus, in the study, we will answer the following question: what are the main difficulties of the emergency transition to distance learning and what changes need to be made to the presentation of the course "Methods of Teaching Mathematics" to overcome them? The purpose of the study is to identify the features of preparing and conducting online classes that contribute to solving the problems of introducing a new subject and achieving the planned educational results in the discipline.

2 Survey Methodology

The implementation of distance learning in the MPSU was carried out using the electronic information and educational environment InfoDa (LMS Moodle).

When creating the course, a systematic approach was used, focusing on establishing the interrelationships of all components of the educational process. In the context of distance learning, it is important to provide students with not just a fragment of educational information, but a complete educational product, including goals, content, forms and methods, and learning outcomes.

Classes began in-person and involved lectures (once every two weeks) and practical exercises (weekly). When switching to a distance format, it was necessary to propose a new form of interaction and to design training and evaluation materials for an electronic course "Methods of teaching mathematics" in the e-learning environment of the Moscow State Pedagogical University InfoDa (LMS Moodle).

2.1 Prerequisites for the Search for New Forms of Education

The experience of conducting the first lessons in the mode of videoconferencing showed that students were not ready to perceive educational information on the discipline within 90 min. in the traditional form that was adopted during the classroom lecture. In the absence of personal contact, students behaved passively: they were distracted by extraneous activities, were not ready for a short discussion, did not enter into a dialogue, did not show initiative to answer current questions during the lecture, where it was supposed to use a problematic style of presentation.

This was also due to the technical features of conducting an online lesson, such as the impossibility of a quick reaction to a teacher's question: to answer the question, it was needed to text in the chat, which took a certain amount of time, or to turn on the microphone. If several students did this at the same time, the sound quality did not allow to perceive their answers. Later, the students began to use the "raised hand" signal to answer. In addition, some of the students participated in the lesson using a smartphone, which did not allow them to see the presentation accompanying the lecture in the required quality. Thus, the material planned for one classroom lecture could not be provided in an online format in full and with adequate quality.

The observations led to the need to search for other forms of presentation of educational information and interaction with students related to the use of contextual learning methods that consider the methods of engagement in online learning [4].

2.2 *Organization of the Study of the Theoretical Part of the Course*

As a result, the following approach to mastering the necessary educational information seems appropriate. 90 min of the lecture is distributed as follows:

(1) 15 min.—setting goals for the lesson;
(2) 60 min.—independent work with educational materials in an electronic course posted in InfoDa (LMS Moodle);
(3) 15 min.—current control and answers to students' questions.

In the electronic course, the title of this fragment is provided with a link to educational materials on the topic, as well as an audio recording of the teacher's comment. Each paragraph I-IV is supplemented with a hyperlink to the corresponding electronic resource. Moreover, only the parts of regulatory documents necessary for studying are attached to paragraphs I and II. There is a link to the video fragment of the lesson and its summary to paragraph IV.

The advantage of working with such a structured text is the ability to concentrate on studying the topic and do it in the allotted study time. The student does not waste time on an independent search for the necessary information in the proposed source, it has already been prepared.

2.3 *Organization of a Practical Online Lesson*

The organization of a practical online lesson also had a few peculiarities. Students were offered assignments to create and present their own educational products (EP), as it was planned for classroom learning. However, during classroom learning, criteria and indicators were developed for organizing joint assessment of the created product. Under the prevailing conditions of conducting a lesson in videoconference mode, ready-made universal criteria and indicators were proposed to save study time.

The criteria were formulated taking into account the need to form a few universal, general professional and professional competencies of students in accordance with the discipline program and the Federal State Educational Standard of Higher Education adopted in the Russian Federation (FSES HE) [5]. Indicators and rating points are assigned to each criterion.

It is possible to create a similar survey in LMS Moodle. In the conditions of an emergency transition to distance learning, some of the educational materials were created with the participation of active students and outside the course. Then links to the created materials were placed in the electronic course. This is one of those cases.

3 Results and Discussion

The credit for the discipline was set on the basis of the rating points scored on a dichotomous scale and in the presence of a portfolio of own educational products created in the course of studying the discipline. Of the 75 students, 63 people scored enough points to receive credit, which made up 84% of the total number of students. The rest passed the test after completing additional compensating tasks. We believe that such rather high learning outcomes do not fully reflect the mastery of the required competencies. This is also due to the conditions for completing assignments. The current tasks in the test form could be completed by students independently, and the creation of their own educational products was carried out in small groups. Not every member of the group made the same contribution to the collaboration. The justification of working in small groups is caused by the desire to increase the involvement of students in the educational process through mutual responsibility for the presented EP.

3.1 Difficulties of Emergency Transition to Distance Learning

Let us systematize the main difficulties that were identified during the emergency transition to distance learning in relation to the discipline being taught.

Technical and organizational difficulties.

1. Late connection of students to a lecture or absence. Solution: providing a recording of the first stage of the lesson, the late or absent student has to pass the second and third stages on the corresponding section of the e-course with a decrease in rating points. Students log in to the videoconference platform from a smartphone, formal attendance at the lesson. Solution: students connect with video at the beginning and during the lesson.
2. Late submission of completed assignments and the lack of independent fulfilment. Solution: in the electronic course environment, it is possible to limit the time allotted for completing the task. The rating score is reduced for delayed submission of tasks. The control of independent performing of test tasks in a remote format does not seem to be possible without attracting additional means and capacity. Tasks for creating your own educational products cannot be completed independently. They are provided after discussion and assessment in the lesson and subsequent adjustment with the possibility of obtaining a higher grade, which serves as a stimulating factor for the timely submission of completed assignments.
3. A large number of tasks requiring verification and assessment by a teacher. Solution: creating tasks in a test form with the possibility of automated verification, organization of collective assessment of EP created by students during the lesson.

Substantive difficulties.

1. Insufficient preparation of students in pedagogy, psychology, theoretical foundations of the mathematics school course. Solution: identifying knowledge gaps in the field of pedagogy and psychology when performing test tasks and creating together with the students a reference dictionary as an element of the e-course. Implementation of individual tasks on the logical and mathematical analysis of the topic where poor knowledge was found, conducting methodological and mathematical dictations.
2. Unreadiness of students to work independently with scientific and methodological text and to create their own educational products. Solution: structuring of the educational information, presenting it concisely with hyperlinks first by a teacher and then on their instructions by students. Completing assignments for a practical lesson in small groups.
3. The need for a teacher to prepare training information, assignments and other materials for posting in e-course in a short time. Solution: using training materials requiring minimal revision, reducing the requirements for content design, attracting assistants from among the students.

Most of the identified problems also occur in offline learning. However, when working in information and educational environment, all student actions are automatically recorded, intermediate and final rating points are calculated, which allows both a teacher and a student to receive operational information about the progress in the study of the course. In addition, a student has permanent access to training course information required and sufficient for its mastering in accordance with requirements of the working program. All this is an undoubted advantage of distance learning.

3.2 Results of the Assessment of the Residual Knowledge of Students

In September 2020/21 the residual knowledge on the theoretical part of the studied course of now fourth-year students was assessed in offline format. For several topics of the course, 14 test tasks were offered in a test form.

The assessment results showed that 54 out of 75 students scored enough points for a positive mark, which amounted to 72% of the total number of students. Compared to the results obtained on the test, the percentage of students who received a positive grade decreased by 12%. Such a decrease is natural and acceptable.

The analysis of the completed tasks showed that students know special terminology at the level of recognition but make mistakes when using it in the educational text. This, in particular, indicates insufficient experience in working with methodological and mathematical information and sets tasks for subsequent training in this discipline.

For the current assessment of the assimilation of educational information, two types of tasks were also used according to two levels of assimilation of educational material. When performing tasks of the first type, reproduction of information was required, in tasks of the second type it was necessary to demonstrate an understanding of what had been studied and the ability to apply knowledge in a familiar situation. These types of tasks were used in online lecture classes for current control (a total of 14 tasks of the first type and 16 tasks of the second type).

As a result of comparing the success of students' performance of tasks of each type, it was found that during the current control, more correctly completed tasks turned out to be among the tasks of the first type. The test of residual knowledge showed that students completed the second type of tasks more successfully. This is illustrated in Figs. 1 and 2. The number of the task is indicated horizontally, and the number of students who correctly solved the task is indicated vertically.

Thus, if in the course of training students more successfully completed tasks of the first type, then after completing the course, students reproduced the required information inaccurately or with insufficient completeness in tasks of the first type but demonstrated the ability to use it when performing tasks of the second type.

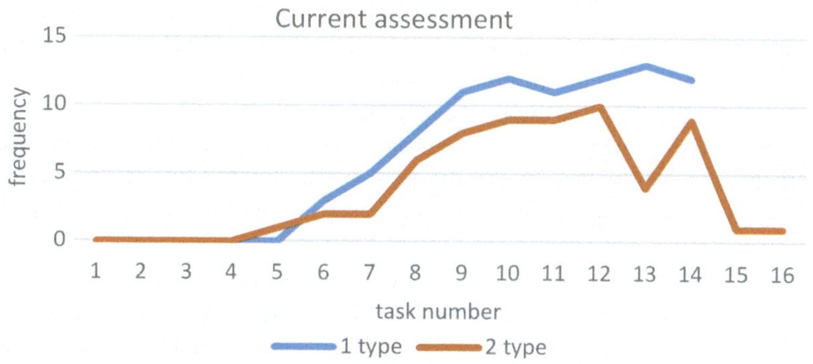

Fig. 1 Comparison of the performance of two types of tasks under the current control

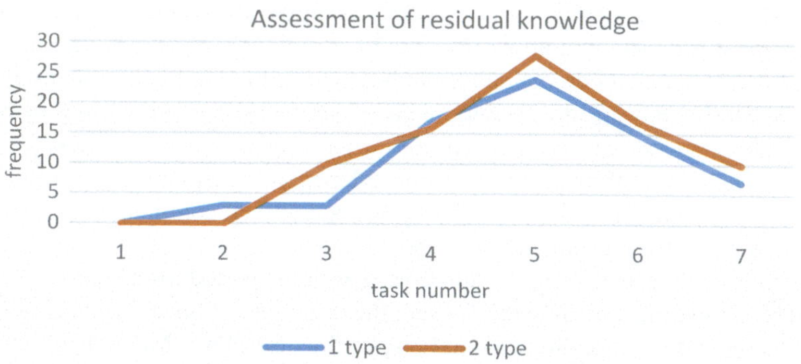

Fig. 2 Comparison of the performance of two types of tasks when checking residual knowledge

3.3 *Deficiencies in Methodological Training of Students*

Despite the identified positive trend towards improving the quality of assimilation of educational information, own experience of supervising coursework on the methodology of teaching mathematics in the fourth year and conversations with colleagues showed that students are not well enough oriented in the methodological literature, approach the compilation of sets of tasks for teaching school children formally, don't show independence and enthusiasm when conducting research. At the same time, it was noted that when students face certain difficulties, they turn to the lecture materials of the course and study not only the main content of the lecture, but also follow most of the proposed links. We consider this as a positive result.

4 Conclusion

Thus, the transition to a distance learning format for 3rd year bachelor students made it possible to identify a few difficulties in the study of the discipline "Methods of teaching mathematics" and propose ways to overcome them.

It was also possible to fill the electronic course with educational information presented in a form that allows to study the discipline "Methods of teaching mathematics", and to refer to it in the future when preparing educational research papers.

The experience of interacting with students in a distance and offline format allows us to claim, following several researchers [1], that the quality of distance learning is somewhat lower. Further search for forms and means is required to motivate students to work independently on the study of the discipline necessary for any format of the educational process, ways of interaction between students and the teacher, improving the content of the control and measuring materials of the course for current and final control. The expansion of topics and levels of complexity of tasks for the development of educational products will allow students to consciously choose in the future the topics of their own methodological research conducted in term papers and theses.

As the study showed, the development of strategies for presenting an online educational course should be focused not on the transfer of information from the teacher to the student through the Learning Management System, but on the organization of interactive learning, which implies operational communication between all participants in the educational process.

It should be noted that during the subsequent training period from February to June 2021, which took place in a mixed format, the materials of the created course have greatly facilitated the teaching and studying of the discipline "Methods of teaching mathematics" for new third-year students. In addition, the work of students in the educational environment contributed to the improvement of their professional skills.

References

1. Karakozov SD, Kovalev EE, Manyakhina VG, Muravyeva OV, Nikiforova AV, Smotryaeva KS (2021) Problems and results of forced transition to distance learning of students and teachers (Institute of mathematics and informatics of Moscow state university). Teach XXI cent 1(1):11–23
2. Saidi RM, Sharip AA, Abd Rahim NZ, Zulkifli ZA, Md Zain SM (2021) Evaluating students preferences of open and distance learning (Odl) tools. Proc Comp Sci 179:955–961
3. Emmerichs L, Welter VDE, Schlüter K (2021) University teacher students learning in times of COVID-19. Ed Sci 11:776
4. Kwon S, Kim W, Bae C et al (2021) The identity changes in online learning and teaching: instructors, learners, and learning management systems. Int J Ed Technol High Ed 18:67
5. Yavuzalp N, Bahcivan E (2021) A structural equation modeling analysis of relationships among university students' readiness for e-learning, self-regulation skills, satisfaction, and academic achievement. RPTEL 16:15

Digitalization of Mathematics Education at a Technical University Need and Reality Amid the Pandemic

Svetlana Rozanova, Alexandra Potepalova, and Irina Dragileva

Abstract The paper is devoted to the study of the impact of distance learning on the quality of mathematics education at technical universities in the context of digitalization amid the pandemic. The analytical review of the literature and the pedagogical experience of the authors made it possible to define the problem of the study: to identify the impact of distance learning on the quality of mathematics education at a technical university in the context of digitalization amid the pandemic. The relevance of the study is justified by the need and reality of the digitalization of mathematics education in universities in the conditions of the pandemic. The purpose of the study is to develop and implement a conceptual and methodological complex aimed at reducing the negative aspects of distance learning in mathematics at a technical university and maintaining the quality of mathematics education amid the pandemic. Research hypothesis: the use of the positive aspects of digitalization and the implementation of the developed conceptual and methodological complex in the transformation of the educational process in mathematics at a technical university will contribute to its quality amid the pandemic. The conceptual and methodological complex implemented in the mathematics education based on the obtained results of the experiment allows concluding with a probability of 0.95 that the level of students' performance in mathematics is maintained in the conditions of distance learning during the pandemic.

Keywords Digitalization in education · Mathematics education · Pandemic · Distance learning · Technical universities · Conceptual and methodological complex

---

S. Rozanova (✉) · A. Potepalova · I. Dragileva
MIREA—Russian Technological University, 78, Vernadsky Ave, Moscow 119454, Russia
e-mail: srozanova@mail.ru

A. L. Semenov et al. (eds.), *Structural and Technological Transformation of Education in the Post-Pandemic Period*, Lecture Notes in Networks and Systems 1282,
https://doi.org/10.1007/978-3-031-84039-5_9

1 Introduction

1.1 Relevance of the Study

The relevance of this study is justified by the need and urgency of the digitalization of mathematics education in universities, which is accelerated by the pandemic. The need to digitalize higher education is caused by global changes in the global economy from the late twentieth century, which further accelerated in the twenty-first century. Since 2017, a number of fundamental documents on the digitalization of secondary and higher education have appeared in Russia. But from the beginning of 2020, the pandemic accelerated the need for digitalization and amplified its problems.

1.2 Literature Review

An analytical review of foreign literature allows us to highlight a number of areas of research related to the digitalization of higher education, accelerated by the pandemic. An in-depth analysis of works in some areas is given in our article [1].

Let us discuss the next direction of study in more detail. It is the works that are of particular interest to the topic of this article: works exploring the pros and cons of digitalization in education; devoted to the impact of distance learning during the pandemic on various characteristics of the educational process, mainly mathematics (motivation of students and teachers, role of new technologies, academic progress, education management).

The works identifying and updating the problem topics and problem areas in teaching mathematics at universities began to appear in 2011 but peaked in 2017–2018 [2–4]. This work was necessary due to the existing gap in the mathematical knowledge of applicants and the requirements of technical universities for their mathematical training. The complexity of this problem is intensified by the lack of motivation for many applicants to technical universities to study mathematics, greater development of technical thinking than mathematical. Studying higher mathematics makes such students accumulate those problem areas.

The researchers note the following:

– The pandemic has become a global problem of education. The pandemic accelerated the process of digitalization in universities.

– Various research methods are used in the processing of experimental data: methods of descriptive statistics [5], correlation and regression analysis, cluster analysis, fuzzy logic methods [6].

Open issues:

– Insufficient attention is paid in pedagogical study of the impact of the pandemic on the quality of mathematics education to the specifics of the university profile.

– The study of higher mathematics, especially during the pandemic, leads to the fact that students not motivated to study mathematics accumulate problem areas through learning teaching higher mathematics, which significantly affects the quality of the educational process.

– Attention is not focused on the search for conceptual ideas, methods and means that minimize the negative impact of online education in the pandemic.

1.3 Problem Statement

Many years of experience of the authors of this article in teaching higher mathematics at a technical university and these unresolved issues made it possible to formulate the problem and the purpose of study.

To achieve this goal and solve the problem of study, the tasks of the study are set as follows:–

– To improve the pool of problem areas of basic educational and research elements of higher mathematics, methods and algorithms of solving complex problems associated with the problem area in the conditions of existing reality.

– To develop and implement a conceptual and methodological complex aimed at reducing the negative impact of distance learning in mathematics at a technical university during the pandemic.

– To set the problem and ensure statistical processing of results of the experiment to identify the impact of distance learning in mathematics on the quality of mathematics education at a technical university in the conditions of the pandemic and to implement the developed conceptual and methodological complex.

2 Research Methodology

2.1 Theoretical and Methodological Framework of the Study

The paper is devoted to the development and implementation of a conceptual and methodological complex aimed at reducing the negative impact of distance learning in mathematics at a technical university and maintaining the quality of mathematics education during the pandemic. This determined the preliminary development of the theoretical foundations of the methodology and, as a result, its methodology.

The methodological unity consists of five conceptual ideas, on the basis of which it is advisable to perform a gradual transformation of the academic process in higher mathematics at technical universities in the conditions of digitalization of education:

1. Reasonable digitalization of mathematics education at a technical university.
2. Synthesis of traditional and latest methods and techniques.
3. Preservation and implementation of the principle of the optimal combination of fundamentality and applied focus in the educational process in mathematics at a technical university in the context of education digitalization.
4. Integration of mathematical, informational, professional and general cultural knowledge.
5. Attracting students to research work from the first courses of study.

2.2 Methodology for Using the Potential of Problem Areas

The methodology for using the potential to disclose the problem areas in the transformation of the educational process in mathematics in the context of digitalization amid the pandemic is outlined in our article [1].

2.3 Research Framework

The research framework is considered in this article on the example of adjusting the elements of the educational process at the Department of Higher Mathematics of MIREA—Russian Technological University. The digitalization of the university caused the development and introduction of the educational environment—MOODLE (Modular Object-Oriented Dynamic Learning Environment)—a "dynamic learning environment". This system includes areas dedicated to all areas of mathematics.

2.4 Participant (Subject) Characteristics

An experiment was prepared and conducted at the Department of Higher Mathematics of MIREA—Russian Technological University to study the impact of distance learning in mathematics during the pandemic on students' progress provided the developed conceptual and methodological complex is used. Two groups of students of the specialist mode of study of one of the areas of study were randomly selected for the experiment. Those students were enrolled in different years (2018, 2019, 2020) with the same basic program in mathematical analysis. The progress in mathematical analysis in the 1st, 2nd and 3rd semesters was considered. The control group included students of two groups in the amount $N = 56$ during all three semesters that began their studies in 2018. The considered period of study for them was before the pandemic, but in the conditions of digitalization that began in 2017. The experimental group included students of two groups in the amount M = 56 (first semester), M = 60 (second and third semesters) who joined the technical university in 2019. They began to study offline before the restrictive measures introduced in 2020, but they had to study online during the 2nd and 3rd semesters. Lectures and practical classes were held remotely in the form of webinars, control works—remotely, exams—in the form of tests. They were deprived of face-to-face contact with the teacher. Besides, students of two groups in the amount M = 46 (first semester), M = 44 (second semester), M = 43 (third semester) enrolled in 2020 and studying all three semesters during the pandemic were taken as the second experimental group. Data for statistical calculations were taken according to the final assessment reports, i.e., the results of exams and retaking of unsatisfactory grades for positive grades.

2.5 *Study Procedure*

The structure of a conceptual and methodological complex was developed to solve the tasks of the study. The structure of the complex includes the following components.

1. Five conceptual ideas.
2. Methods of using methodological and organizational possibilities of MOODLE.
3. Lectures, seminars and teaching aids in electronic form.
4. Methodology for using the potential of disclosing the problem areas.
5. Pool of identified mathematical problem areas with tables and algorithms of their disclosure; pool of professional and applied tasks; list of topics of research works for students of the specialist mode of study and topics of graduation works for bachelor students.

The theoretical assumption of the authors on the efficiency of using the complex in the educational process in mathematics at a technical university, especially during the pandemic, was confirmed by statistical processing of the experimental results the use of this complex helped preserving the level of student progress. Table 1 presents the summary of earlier identified problem areas from various sections of mathematics. In order to save the length of the article, no more than two examples of "problem areas" are presented for each section of higher mathematics.

Table 1 Problem areas within the sections of higher mathematics

Section of higher mathematics at technical universities	Topics that require special teaching approaches	What methods were used: innovative or/and traditional
Mathematical analysis, 1st semester	–Limit of numerical sequence and the function of one variable –Seven types of uncertainty: $\frac{0}{0}, \frac{\infty}{\infty}, 0 \cdot \infty, \infty - \infty, 1^{\infty}, 0^{0}, \infty^{0}$.	– isualization using CT, programs, leading the idea of mathematical analysis –Algorithms, diagrams, elements of business game
Mathematical analysis, 2nd semester	–Definite Riemann integral over domain	–Use of physical and geometric meaning, algorithms, schemes, visualization (illustrations) using CT, application to problems of physics
Mathematical analysis, 3rd semester	– Approximation of the periodic function by partial sums of the Fourier series	–Algorithm, step-by-step disclosure of problem areas, programs, CT, applications, individual-dual electronic reviews

3 Results and Discussion

3.1 Statistical Evaluation of Results

Diagrams 1, 2, 3 compare the academic performance of the control group (2018), experimental (2019) and experimental (2020) groups (Figs. 1, 2 and 3).

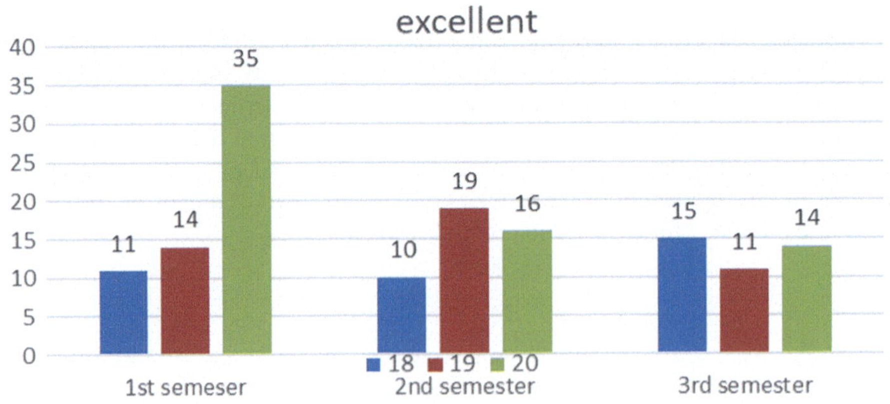

Fig. 1 Comparison of the academic performance of the control group and experimental-groups (2019 and 2020) according to the "excellent" parameter

Fig. 2 Comparison of the academic performance of the control group and experimental groups (2019 and 2020) according to the "good" parameter

Fig. 3 Comparison of the academic performance of the control group and experimental groups (2019 and 2020) according to the "satisfactory" parameter

The above diagrams show that before the experiment (1st semester), the academic performance of the control and experimental group 2019 were almost the same.

The academic performance was different by the end of the period under review. In the second semester, the experimental group (2019) performed better than the control group (2018). In the third semester the differences were "the other way", but they were insignificant. The level of differences in experimental data was determined using the Cramer-Welch test (Student). The calculation according to this criterion on the equality of averages, 2018 and 2019: $T_{Imp.} = \frac{\sqrt{M \cdot N}|\overline{x}-\overline{y}|}{\sqrt{M \cdot D_x + N \cdot D_y}}$, if 2018 is: $(x_1, \ldots, x_N)$, 2019: $(y_1, \ldots, y_M)$. **Semester 1**: $T_{Imp.} = 0.049$ **Semester 2**: $T_{Imp.} = 1.411$;

Semester 3: $T_{Imp.} = 0.901$. All numbers do not exceed $T_{0.05} = 1.96$. Conclusion: the characteristics of the compared samples coincide with the confidence level $P = 0.95$. The academic performance of the control group (2018) was also compared with another experimental group (2020), the entire study period of which fell on the pandemic. Seminars were held in a mixed format, then offline, then remotely, and all lectures—in the form of webinars. In the experimental group (2020), similar to 2019, the conceptual and methodological complex was used. In this case the group-2018 cannot be called the control group in its strict sense, since according to the initial data of the performance in mathematics, the groups 2018 and 2020 are distinguishable, but some interesting conclusions from such a comparison can be obtained.

The diagrams show that by the end of the third semester the differences in the academic performance of the groups were reduced to a minimum. Thus, in the most difficult conditions of the ongoing pandemic, the pedagogical faculty managed to maintain the level of academic performance of students of experimental groups 2019 and 2020.

3.2 *Discussion of Results*

The experiment showed:

- The developed conceptual and methodological complex, its implementation in the educational process in mathematics in the experimental groups of MIREA and the results of statistical processing of experimental data to identify the impact of distance learning during the pandemic on the academic performance of students make it possible to conclude with a probability of 0.95 on the stability of the academic performance of students of the technical university.
- The use of the technique of disclosing the potential of problem areas when teaching higher mathematics to students of technical universities who have more developed technical thinking than mathematical ones help teachers significantly to overcome emerging difficulties, especially during the pandemic;
- The conclusion arising from the experiment on the possibility of preserving academic skills in mathematics among students of a technical university under a remote mode of study in emergency circumstances is consistent with the results of the studies [5–7].

4 Conclusion

The analytical review of the literature, theoretical and practical studies conducted in this article lead to the following conclusions.

1. The conceptual and methodological complex was created, which is described in paragraph 2.5. It takes into account the conditions and methods that minimize the negative impact of the pandemic.
2. The theoretical and practical research with the setting up of an experiment and statistical processing of its results confirmed the hypothesis expressed in the Abstract, helped solve the problem and objectives of the study, and achieve the goal.
3. Creation of better virtual learning environment and web platforms will help high school teachers improve learning methods to disclose the problem areas; push remote lectures, seminars and exams in cases of the pandemic closer to full-time education; serve to form the creative abilities of students and, in the end, to improve the quality of mathematics education in technical universities.

References

1. Dragileva IP, Potepalova AY, Rozanova SA (2021) The need to use the potential of revealing "problem areas" to improve the quality of mathematics education in technical universities in the context of digitalization of higher education. Contin Math Comput Sci Educ 4:109–124. https://doi.org/10.24888/2500-1957-2021-4-109-124
2. Dvoryatkina SN, Dyakina AA, Rozanova SA (2017) Synergy of humanitarian and mathematical knowledge as a pedagogical state of solving interdisciplinary problems. Ed Integrat 21(1):8–18. https://doi.org/10.15507/1991-9468.086.021.201701.008-018
3. Ní Shé C, Mac an Bhaird C, Ní Fhloinn E, O'Shea A (2017) Problematic topics in first-year mathematics: lecturer and student views. Int J Math Ed Sci Technol 48:715–734. https://doi.org/10.1080/0020739X.2016.1272142
4. Smirnov EI (2017) Synergy of the study of the "problem zone" of the basic educational element of the content of mathematics education. Yarosl Pedag Bull 5:82–90
5. Sozontova EA, Prodanova NA, Zekiy AO, Rakhmatullina LV, Konovalova EV (2020) Efficiency of remote technologies on the approaching the content of a e-learning mathematics course using MOODLE system: case study. PeriodicoTcheQuimica 17(35):1159–1174
6. Capone R, Lepore M (2021) From distance learning to integrated digital learning: a fuzzy cognitive analysis focused on engagement, motivation, and participation during covid-19 pandemic. Technol Knowl Learn. https://doi.org/10.1007/s10758-021-09571-w
7. Jeffrey L (2021) Poor performance in undergraduate math: can we blame it on Covid-19 despair? IJISM 9(3):31–40. https://www.ijism.org/administrator/components/com_jresearch/files/publications/IJISM_934_FINAL.pdf

Intelligent Ranking for the Results of Students' Knowledge Control Using Machine Learning Methods

Alexey A. Petrov, Olga V. Druzhinina, and Olga N. Masina

Abstract The article deals with the methods of intelligent ranking of knowledge control results in the context of digital education in mathematics. The article considers approaches to learning using hybrid intelligent learning environment and features of digital transformation of mathematics education. The methods of intellectual ranking of the results of intermediate knowledge control with regard to adaptation to the digital format of education are proposed. Machine learning algorithms are proposed for modelling the educational process, in particular, generalised clustering algorithms based on the k-means method and cascade learning of autoencoder. The developed algorithmic and software is aimed at intelligent evaluation of the results of intermediate knowledge control and identification of research potential. The conducted computational experiments allowed analysing and interpreting the obtained results. The proposed models and algorithms can be used to improve the methods of knowledge control and assessment, as well as to predict the indicators of the pedagogical process.

Keywords Machine learning · Autoencoder · Feature extraction · Clustering · Hybrid intelligent learning environment · Knowledge control and assessment

1 Introduction

Improvement of instrumental and methodological support of the pedagogical process using intelligent technologies is an actual and important task in modern education. In this context, the development of an integrated complex combining knowledge base and artificial intelligence tools for effective assessment of students' knowledge within

A. A. Petrov · O. V. Druzhinina · O. N. Masina (✉)
Bunin Yelets State University, 28, Kommunarov St, Yelets 399770, Russian Federation
e-mail: olga121@inbox.ru

O. V. Druzhinina
Federal Research Center "Computer Science and Control" of Russian Academy of Science, 44, Building 2, Vavilov St, Moscow 119333, Russian Federation

A. L. Semenov et al. (eds.), *Structural and Technological Transformation of Education in the Post-Pandemic Period*, Lecture Notes in Networks and Systems 1282,
https://doi.org/10.1007/978-3-031-84039-5_10

the framework of a hybrid intelligent learning environment is of special theoretical and applied interest. Hybrid Intelligent Learning Environment (HILE) is a modern educational platform that combines traditional teaching methods with innovative artificial intelligence tools [1–3]. This allows for a more effective and personalised learning system. Within the framework of GIOS, special attention is paid to the development of intelligent knowledge assessment systems that can analyse test results, assess the level of mastery of the material and provide feedback to students. In [4], a formalisation of the structure of the knowledge base is proposed, taking into account its use in the HILE module related to the control and assessment of knowledge in mathematics. However, the issues related to the development of new algorithms for intelligent analysis of test results require further study and improvement [5]. The development of new algorithms for the intellectual analysis of test results will make it possible to improve the accuracy and efficiency of students' knowledge assessment, as well as to make the learning process more individualised and effective. Artificial intelligence in education is an actively developing area [6–8]. Aspects of designing individual educational routes using HILE in the context of digitalization of education are considered in [2, 9, 10]. The use of some intelligent technologies and neural network modeling methods in the development of HILE is described in [2, 5, 11, 12]. In particular, the use of methods for solving problems of clustering, dimensionality reduction and classification made it possible to develop the foundations for techniques creating, operational learning, monitoring and evaluating knowledge in mathematics in a hybrid educational environment [4].

Data mining associated with clustering, classification and dimensionality reduction is an important and relevant direction. When solving clustering and classification problems, machine learning methods are effective [13, 14]. Various problems within this direction are considered, in particular, in [15, 16]. Among the relevant software tools, open-source machine learning libraries Scikit-learn, Keras, Tensorflow, Pytorch, Clustering.jl should be noted [17]. The implementation of intelligent algorithms requires the use of advanced mathematical software. The tools for creating such software include the Python and Julia languages with the mathematical libraries [18]. A growing area of data analysis based on unsupervised machine learning is the use of autoencoder networks. The advantage of this approach is the possibility of nonlinear multivariate data analysis with feature extraction. For example, the papers [19–21] are devoted to various aspects of using autoencoders in data analysis. Verification of intelligent models can be carried out using various types of experiments and with the construction of auxiliary probabilistic models [22–25].

In the present paper we consider the problems of intellectual ranking for the results of students' knowledge control using machine learning methods. We are constructing an autoencoder and developing a machine learning algorithm for data clustering within a hybrid intelligent learning environment.

2 Modeling of Hybrid Intellectual Learning Environment

The features of the hybrid intelligent learning environment (HILE) are described in [3–5, 11]. When forming the evaluation materials of the assignment, three blocks are provided (A1–A3), which correspond to the levels "student", "typical" and "heuristic", respectively [4, 22]. In addition, as part of the extension of the methodology, the possibility of adding block B1 is being considered. Block B1 is represented by questions of increased complexity, the answers to which require creative competencies. The tasks of block B1 require development using, in particular, tasks of increased complexity and Olympiad tasks.

In this study, we use such a generalized evaluation scale with updated criteria [4], which, on the one hand, is based on the development of the three-level methodology of V.P. Bespalko, and on the other hand, takes into account the realities of the pedagogical process of the current time. The student demonstrates the following assimilation levels.

1. Unacceptable level of assimilation if less than 70% of the tasks of each block A1, A2 and A3 are correctly completed.
2. Low level of assimilation, if more than 70% of the tasks of at least one of the blocks A1, A2 or A3 are completed correctly.
3. Average level of assimilation, if more than 70% of the tasks of two blocks out of three are correctly completed.
4. High level of assimilation, if more than 70% of the tasks of each of the three blocks A1, A2 and A3 are completed correctly.

The assimilation levels associated with block B1 (increased complexity) are aimed at identifying the research potential of students through expert evaluation.

To evaluate the test results obtained, we use a combined method based on the use of fixed ratings ("unacceptable", "low", "medium", "high") and on the use of an intelligent ranking algorithm. For the specified ranking algorithm, we choose a semi-supervised machine learning. Algorithm 1 consists t of the following steps.

1. Data preprocessing.
2. Data clustering (k-means, c-means, evolution algorithms, etc.).
3. Search for centroids.
4. Marking of centroids using expert assessment.

To model the pedagogical process under study and to test the work of the ranking algorithm, we propose a probabilistic learning model based on quantitative assessments of the competencies level. When constructing this model, we are based on the assumption that the levels of competence development in the field of mathematical disciplines are determined using a normal distribution. To implement a probabilistic model and to conduct computational experiments, a program is developed in the Python 3 language using Numpy, Matplotlib libraries.

3 Results of Computational Experiments

To test the evaluation methodology, computational experiments are carried out using the probabilistic model of learning presented in Sect. 3. For blocks A1–A3, the competence levels determined by the normal distribution in the range from 0 to 1 are considered. The threshold competence levels are 0.5 for blocks A1–A3.

The splitting of the results obtained is carried out in accordance with fixed ratings ("invalid", "low", "medium", "high"), as well as using the ranking algorithm (Algorithm 1) with the corresponding ranks ("rank 1", "rank 2", "rank 3", "rank 4"). To reduce the dimensionality of the resulting vectors, the average value for the block is used. A program is developed in Python using SciPy/NumPy libraries to implement a probabilistic and ranking models. At one of the steps of algorithm 1, the *k*-means ++ algorithm from the Scikit-learn library is used. The results of the model testing of 500 students in relation to the content blocks A1–A3 are shown in Fig. 1.

The breakdown of the results using the ranking method (Fig. 1) is represented by four clusters ("rank 1", "rank 2", "rank 3", "rank 4"), each of which includes a comparable number of results. It is important to note that this partitioning is intended solely to facilitate expert evaluation of the results, since the number of clusters is constant. The breakdown of the test results by the block of questions of increased complexity B1 is shown in Fig. 2.

Using a ranking algorithm based on machine learning, the results are divided into three clusters: "low creativity" (red dots), "medium creativity" (yellow dots), "high creativity" (green dots). The centroids are indicated by the crosshairs in Fig. 2. Unlike

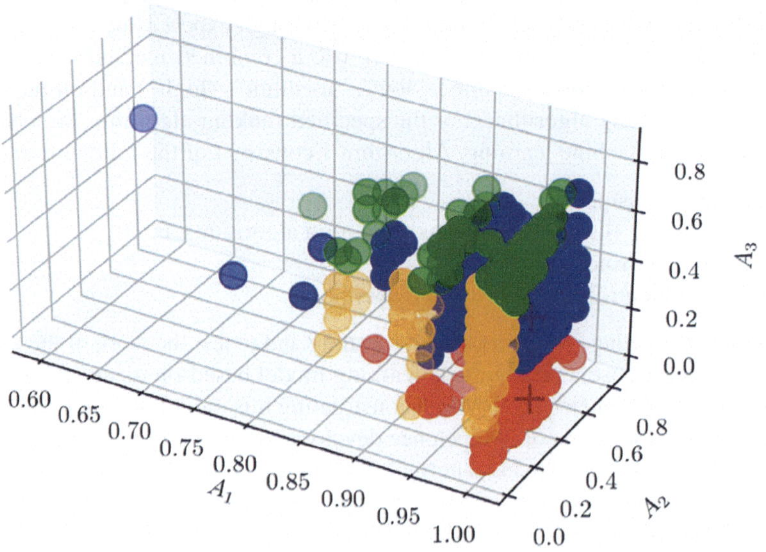

Fig. 1 Splitting the results according to the ranking method using machine learning

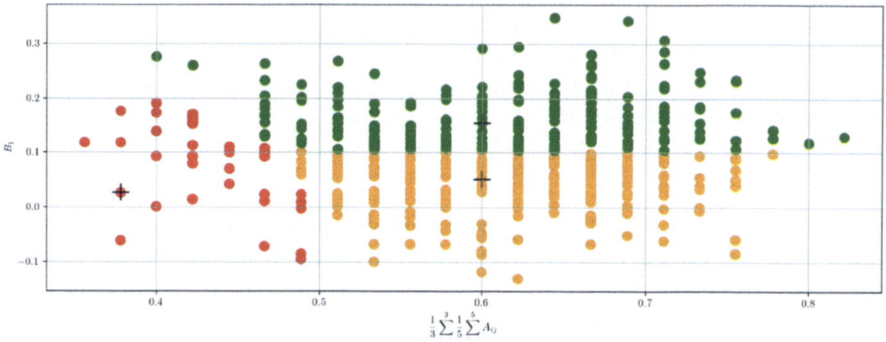

Fig. 2 Test results for block B1 with questions of increased complexity

blocks A1–A3, the value of the competence level is considered for block B1 on an extended scale from −1 to 1. It should be noted that the results obtained by analogy with the division in Fig. 1 are intended for expert evaluation.

It should be noted that the possibility of obtaining the results presented in Figs. 1 and 2 is associated with the use of artificial intelligence methods and can be implemented in the development of intelligent information support systems for the pedagogical process in mass conditions when large amounts of data are analyzed. Based on the implementation of a probabilistic model and ranking, the approach to modeling the pedagogical process allows you to create a sufficiently effective combined tool for evaluating and controlling knowledge.

4 Feature Extraction

During testing, large-dimensional data and sparse data may occur. In such conditions, it is often necessary to use more complex analysis methods compared to cluster analysis. The disadvantage of the clustering algorithm based on *k*-means is the impossibility of in-depth analysis, since this algorithm uses only linear metrics of the distance between vectors. To increase the efficiency of ranking test results, it is advisable to use nonlinear feature extraction algorithms. Among these algorithms, we can mention a family of algorithms using an autoencoder with bottleneck topology.

Next, we will adopt the following terminology. The answer vector is understood as a binary column vector in which 0 and 1 mean the correct and incorrect answer to the task, respectively. The shift vector of the neural network will be called such a vector $s(s_1, s_2, \ldots, s_n)$ which determines the change in the number of neurons in layers.

We propose a cascade algorithm for autoencoder training an (algorithm 2). This algorithm consists of the following steps.

1. Initialization of the autoencoder with the topology $n \to (n-1) \to n$, where n is the dimension of the response vector.
2. Calculation of weight coefficients using an optimization algorithm.
3. If the stop conditions are met, the algorithm ends. Otherwise, go to step 4.
4. Change the network topology based on the shift vector.
5. Go to step 2.

A program in the Julia language is developed to implement algorithm 2. We consider such an autoencoder with one hidden layer. The neurons of the input layer are i_j ($j \in (1, 15)$), the neurons of the hidden layer are h_k ($k \in (1, l)$), the neurons of the output layer are u_j ($j \in (1, 15)$), the biases are i_0, h_0. It should be noted that the number of neurons in the hidden layer changes taking into account the displacement vector.

The activation structure of the autoencoder is defined as follows. For the hidden layer, the DreLU (Dual Rectifier Linear Unit) activation function is used, the Heaviside function is used in the output layer. To train a neural network, we choose an algorithm of differential evolution taking into account the effectiveness of this algorithm in the considered topology of the neural network.

We perform computational experiments in order to estimate such an error (data loss), which occurs due to a decrease in the dimension of the data. The data under consideration includes 20 binary answer correctness matrices.

For comparison with the autoencoder, we use the well-known principal component analysis method (implementation PCA of the MultivariateStats.jl library). A comparative analysis of the results allows us to carry out Table 1.

In Table 1, the first column shows the dimension of the reduced vector space. Columns 2 and 3 show the mean square deviation (for the autoencoder and PCA, respectively), that corresponds to data loss ratio. The obtained results indicate that the autoencoder has an advantage over PCA for dimension n $\leq$ 8.

To conduct computational experiments, a program in the Julia language is developed using the Clustering.jl library. The distribution of results by clusters for the PCA algorithm and for the autoencoder are shown in Figs. 3 and 4, respectively.

Figures 3 and 4 visualize 20 vectors corresponding to the test questions answers. For each of these vectors (x_{1i}, x_{2i}, x_{3i}), where $i = 1, \ldots, 20$, the values of x_{1i}, x_{2i} are represented by the coordinates of the circles, and the value of x_{3i} is represented by the changing sizes of the circles. We mark 4 clusters of answers with the help of color coding and line drawing. It should be noted that the distribution of test results in Figs. 3 and 4 corresponds to the same number of vectors in each of the clusters. This circumstance indicates the presence of repeatability in the answers to test tasks.

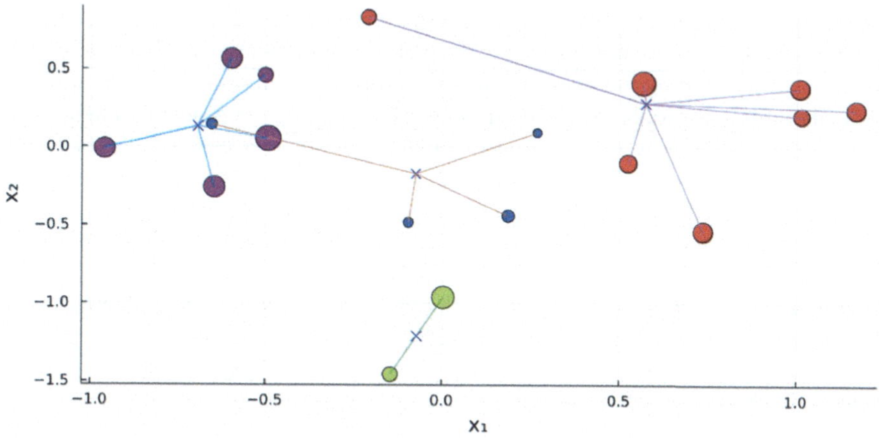

Fig. 3 Distribution of 20 test results across 4 clusters (PCA algorithm)

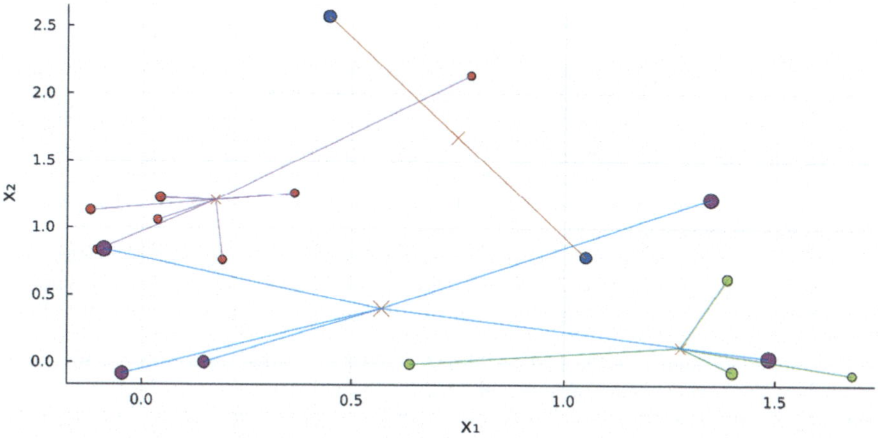

Fig. 4 Distribution of 20 test results across 4 clusters (autoencoder)

5 Conclusion

When processing test results, problems often arise related to the complexity of analyzing multidimensional arrays, in particular, with the complexity of processing sparse and repetitive data. It is convenient to solve such problems with the involvement of such type of artificial neural networks as an autoencoder. In this paper, it is proposed to use an autoencoder for monitoring and evaluating knowledge within the framework of the HILE. The use of an autoencoder allows to reduce the dimension of input vectors with minimal loss of information, with the isolation of sparse data, and also allows to uncover deep correlations of test results. Taking into account these possibilities, the general features and patterns of test data are clarified in comparison with the results of using classical data analysis algorithms. The obtained results are aimed at developing methodological and algorithmic support for the creation of artificial intelligence tools within the framework of solving the problem of assessing knowledge in mathematics. The proposed model of intellectual assessment of knowledge and identification of the research potential of students can be used in teaching mathematical disciplines in the secondary education system. Promising areas of research are the development and implementation of algorithms for generating evaluation and didactic materials using the data analysis about test results. The implementation of a hybrid neural network for multicomponent assessment and identification of the research potential of students can be highlighted as a prospect.

Acknowledgements The reported study was funded by RFBR, project number 19-29-14009.

References

1. Basalin PD, Kulikov DA, Maskina YV (2020) Adapting a hybrid intelligent reinforcement learning environment. Mod Inf Technol IT Educ 16(3):788–798
2. Astachova IF, Kiseleva EI (2020) Hybrid intelligent information system for predicting the number of test tasks. Mod Inf Technol IT Educ 16(4):917–926
3. Smirnov E, Dvoryatkina S, Shcherbatykh S (2021) Technology of development of a hybrid intelligent system of teaching mathematics. In: European proceedings of social and behavioural sciences EpSBS 114:761–769
4. Druzhinina OV, Karpacheva IA, Masina ON, Petrov AA (2021) Development of an integrated complex of knowledge base and tools of expert systems for assessing knowledge of students in mathematics within the framework of a hybrid intelligent learning environment. Int J Educ Inf Technol 15:122–129
5. Petrov AA, Druzhinina OV, Masina ON (2021) Modeling knowledge assessment systems within a hybrid intelligent learning environment. Mod Inf Technol IT Educ 17(1):179–189
6. Cheng X, Sun J, Zarifis A (2020) Artificial intelligence and deep learning in educational technology research and practice. Br J Edu Technol 51(5):1653–1656
7. Holmes W, Bialik M, Fadel C (2019) Artificial intelligence in education. Promises and implications for teaching and learning. The center for curriculum redesign, Boston
8. Borba MC, Askar P, Engelbrecht J, Gadanidis G, Llinares S, Aguilar MS (2016) Blended learning, e-learning and mobile learning in mathematics education. ZDM Math Educ 48:589–610
9. Sampsons D, Karagiannidis C (2002) Personalised learning: educational, technological and standardisation perspective. Interact Educ Multimed 4:24–39
10. Loong EYK, Herbert S (2018) Primary school teachers' use of digital technology in mathematics: the complexities. Math Educ Res J 30:475–498
11. Hadzhikolev E, Yotov K, Trankov M, Hadzhikoleva S (2019) Use of neural networks in assessing knowledge and skills of university students. In: 12th annual international conference of education, research and innovation, pp 7474–7484
12. Masina ON, Petrov AA, Druzhinina OV, Shcherbatykh SV (2021) Methods of intelligent ranking results for the intermediate assessment of knowledge on mathematical disciplines. J Phys Conf Ser 2001:012031
13. Aggarwal C (2019) Neural networks and deep learning. Springer international publishing

14. Dulhare UN, Bin Ahmad KA, Ahmad K (2020) Machine learning and big data: concepts, algorithms, tools and applications. Wiley Publishing
15. Filipovych R, Resnick SM, Davatzikos C (2011) Semi-supervised cluster analysis of imaging data. Neuroimage 54(3):2185–2197
16. Ziv B-J, Gifford DK, Jaakkola TS (2001) Fast optimal leaf ordering for hierarchical clustering. Bioinformatics 17:22–29
17. Geron A (2019) Hands-on machine learning with scikit-learn, keras, and tensorflow: concepts, tools, and techniques to build intelligent systems. OReilly
18. Mckinney W (2017) Python for data analysis: data wrangling with pandas, numpy, and ipython, 2nd edn. O'Reilly
19. Ballé J, Laparra V, Simoncelli EP (2016) End-to-end optimized image compression. In: 2016 picture coding symposium (PCS), IEEE, pp 1–5
20. Ribeiro M, Lazzaretti AE, Lopes HS (2008) A study of deep convolutional auto-encoders for anomaly detection in videos. Pattern Recogn Lett 105:13–22
21. Smirnov E, Dvoryatkina S, Martyushev N, Shcherbatykh S (2023) Software package to support students' research activities in the hybrid intellectual environment of mathematics teaching. Mathematics 11(4):952
22. Salha-Galvan G, Limnios S, Hennequin R, Tran V, Vazirgiannis M (2019) Gravity-inspired graph autoencoders for directed link prediction. In: Proceedings of the 28th ACM international conference on information and knowledge management
23. Bespalko VP (2012) Typical pedagogical errors of testing in education. Sch Technol 5:3–11
24. Martyushev N, Shutaleva A, Malushko E, Nikonova Z, Savchenko I (2021) Online communication tools in teaching foreign languages for education sustainability. Sustainability (Switzerland) 13(19):11127
25. Shutaleva A, Starostin A, Nikonova Z, Savchenko I (2022) Migration potential of students and development of human capital. Educ Sci 12(5):324

Teaching Mathematics and Computer Science at the University Under the Coronavirus Epidemiological Situation

Nikolai Puchkov, Tatiana Dorokhova, and Marina Makeeva

Abstract This article is aimed to study the didactic foundations of teaching mathematics and computer science in the context of the epidemiological situation associated with COVID-19 and the limited opportunities for traditional contact work with students. This study is due to the fact that educational services impact the transformation of students' social values in the face of limited face-to-face work with teachers. In the current situation, it becomes necessary to develop specialized educational programs for the formation of both professional and information-mathematical competencies in order to maintain the quality of education. These programs are characterized by maximum possible variability, allowing the creation of individual educational learning trajectories. The article shows that the construction of these programs is based on the idea of a rational combination of the principles and technologies of traditional and digital didactics and these programs are directed at increasing the social activity of students in the conditions of their self-isolation. The proposed methodology is successfully implemented at the Tambov State Technical University (TSTU).

Keywords Quality of education · Distance learning · Digitalization of education · Didactic foundation · Educational programs

1 Introduction

Education during the period of mutual isolation of teachers and learners, caused by the COVID-19 pandemic, has demonstrated a variety of digital pedagogical tools and technologies that enhance the global digitalization of education associated with a total transition to distance learning During the ongoing COVID-19 pandemic, distance learning has become the main technology for transferring knowledge and experience to students around the world [1]. This circumstance served as a guideline

N. Puchkov · T. Dorokhova · M. Makeeva (✉)
Tambov State Technical University, 1 Leningradskaya St, Tambov 392036, Russia
e-mail: marnikma@inbox.ru

A. L. Semenov et al. (eds.), *Structural and Technological Transformation of Education in the Post-Pandemic Period*, Lecture Notes in Networks and Systems 1282,
https://doi.org/10.1007/978-3-031-84039-5_11

for this study, aimed at substantiating the new didactics of teaching mathematics and computer science at the university in conditions of limited contact work with students. In the digital educational the development of the system of students' social values environment is at the forefront.

Knowledge in the life of any person is an asset that is subject to constant renewal, in view of its intensive obsolescence. The education system is faced with the task of not only developing professional competencies, but also transferring cultural values and traditions to the younger generation, educating a harmoniously developed and socially responsible personality based on spiritual values, historical and cultural traditions. If earlier, with the traditional format of education, the process of becoming specialists went through several stages of socialization, now the pedagogical community is faced with the fact of the formation of professional competencies of future specialists in a virtual environment [2]. This gave rise to a specific challenge to pedagogical practice, which requires prompt comprehension and confirms the relevance of this study.

The global education digitalization leads to the creation of a new education didactics, to the development of new educational programs, to the formation of both professional competencies and methods meeting the principles of virtual pedagogy. In new conditions the role of a teacher has changed, if previously he clearly performed the socially significant axiological function of a bearer of values, a keeper of traditions, customs of society and its culture, than in the new educational environment he almost fails to perform it, due to the lack of traditional contact with students. Nevertheless, pedagogical practice shows that the formation of competencies, cultural norms of behavior requires socio-pedagogical conditions for direct communication with students and the emotional coloring of each educational situation, the transfer of socially accumulated professional experience. All countries clearly felt that the education system was not ready for learning in the conditions of limited opportunities for contact work with students without active communication and an emotionally colored component of the educational process. As a result, all the features and problems of transition to a distance learning format have not yet been sufficiently fully defined, the positive and negative aspects of the global education digitalization have not been fully disclosed, the contribution of digital services involved in educational activities, as well as their impact on personality transformation and shaping of social values has not been fully identified [3].

The purpose of this study is to identify the didactic foundations of teaching mathematics and computer science and study the impact of new educational services on the transformation of students' social values in the face of limited face to face work with teachers. It is assumed that in order to achieve the goal we need to:

- Investigate Russian and foreign experience of professional competencies formation, the system of individual social values specifics in the process of professional development in the context of global digitalization of education;
- Develop specialized educational programs for the formation of competencies at a level ensuring such a modernization of existing educational programs that allows

maximum use of the distance learning possibilities and, first of all, enables the construction of individual educational trajectories of a variable nature;
- Implement close, productive interaction between teachers and students in order to overcome the negative aspects of the intensive education digitalization.

2 Survey Methodology

The research methodology is based on the scientific theories of the professional development of future specialists, the principles of changing their worldview, expanding the horizons of knowledge, depending on the conditions dictated by modernity. The tasks of innovative development of enterprises-consumers of specialists in the context of increasing competition predetermine the content of professional competencies. As we found out earlier [4], their formation will be facilitated to a greater extent by information and mathematical competencies related to the problems of mathematical modeling of production processes, the design of various kinds of fundamentally new technological devices, the collection, processing, analysis and systematization of scientific and technical information. The formation of such mathematical competencies of an engineer as the possession of methods of analysis and synthesis of the processes and phenomena under study, the willingness to acquire new mathematical knowledge, the ability to compose mathematical models of typical professional problems and find ways to solve them, to interpret the professional (physical) meaning of the obtained mathematical result are inextricably linked with the problem of developing their information support. Therefore, it is more appropriate to talk about information and mathematical competencies as indicators of achieving the effectiveness of professional competencies, in particular, specialized professional competencies.

The mutual isolation of educational process participants during the pandemic has significantly changed the nature of practical interaction between students and teachers and excluded direct communication in the learning process, depriving the learning process of emotional coloring. Therefore, the process of forming professional competencies in the context of limited opportunities for contact work with students is not possible without studying the impact of digital educational services on the transformation of students' social values in new socio-economic conditions, limited by preventive measures against the pandemic spread.

A comparative analysis of the results of scientific work on the formation of professional competencies in the context of the universal digitalization of education and the formation of social values among young people has revealed various aspects of research areas leading to the above goal. We would like to single out studies aimed at fighting for the quality of education (teaching, learning), at the introduction of standards that guarantee the achievement of the desired results and, first of all, useful for the students themselves. In mathematics, it is not so much the results of calculations, but the possession of the methodology for obtaining them, the identification and promotion of ideas and achievements of the principles of mathematics. Teachers should give credit to known higher-order results (outside the scope of the educational program), planned and projected, to encourage students to carry out personal research. Education should not be based on the results of only closed skills and competencies that do not guarantee the quality of a genuine university education.

In connection with the formation and development of digital didactics, it is necessary to reveal the role and significance of media educators strengthening their position and influence, promoting the process of personality development with the help and on the materials of mass media in order to form a culture of creative communication skills, the ability to interact with the media space [5]. We should explicitly address the creative possibilities of digital technologies and the problems that they present in order to turn the learning process into the most accessible form for mastering and transferring it to a higher quality level [6].

At present we can feel the fact that as a result of the digital revolution, the world undergoes reverse transformation: education, which was previously considered a public good with equal access for all, is now sold to those who can afford specialized services, namely their own conditions. In addition, the forms of distance learning accompanying digitalization are not always absolutely effective in ensuring the quality of education, therefore, an analysis of the factors that motivate students to attend traditional classes, in particular theoretical lectures, which have undeniable advantages, deserves attention. It is noted [7] that this depends on the didactic, communicative and personal competencies of the teacher: the ability to visually connect theory with practice, to present, plan and organize a lecture in an accessible way, to encourage students for success, demonstrating a friendly attitude towards students, supporting feedback. The principles and technologies of media education are very effective (in particular, media game-based learning).

The methodology of the teacher's work in the context of the integration of digital and non-digital resources is changing significantly: the transition from the system "active teacher—passive student" to a system where the teacher is an intermediary between the active student and the knowledge to be mastered is being stimulated. It is shown that the integration of educational technologies increases the level of technological effectiveness of education in terms of its standardization, unification of ensuring continuity, and increases the motivation of students to use the virtual educational environment. E-learning is moving into the category of educational disciplines [8].

The carried out analysis makes it possible to improve the methodology of teaching academic disciplines at a university with less effort when implementing education digitalization programs in conditions of limited contact work with students. First of all, this concerns teaching mathematics and computer science methodology, since the formation of professional competencies in the specialists' training of is largely based on information and mathematical competencies. At the same time, one should take into account the fact that recently there has been a change in the paradigm of values among students from an individual-personal to a paradigm of social orientation, when tolerance, openness, public recognition through social networks, self-development and contribution to society become key [9].

An analysis of pedagogical experience in the field of digitalization of education led to the conclusion that the formation of information and mathematical competence, becomes more successful only taking into account the importance of the system of social values of students in the digital educational environment [10]. This is achieved through the substantiation of the ideas of digital didactics and includes a set of modern digital tools and technologies, selected to solve a particular problem and develop a system of social students' values.

3 Results and Discussion

The advent of digital technologies in the education system during the period of self-isolation requires the availability of information technology equipment. Technical and technological support for distance learning involves new thinking in the virtual space; demands not just the technological development of civilization, but also the emergence of new educational programs, platforms of thinking and forms of individual socialization.

The experience of our work teaching mathematics and informatics at the Tambov State Technical University [11] shows that the solution of the above complex problem is most effectively carried out in terms of ensuring a rational combination of traditional analytical methods of classical mathematics and the promising advantages of digital information technologies in the framework of the formation of information and mathematical competencies. For the purpose of interdisciplinary integration (computer science and mathematics) and the formation of information and mathematical competence, in our opinion, the process of designing specialized educational programs becomes obligatory. Thus, it includes:

- Organization of a creative group of teachers of mathematics and informatics;
- Analysis of the academic discipline work program, identification of similar topics and didactic units available for integration;
- Construction of each integrative lesson.

The concept of "integration" of information and mathematical modules into the general course of higher mathematics must certainly take into account the content of the theoretical part: mathematical concepts, axioms, definitions, theorems in accordance with the identified practice orientation; for one or another area of undergraduate training, these can be noticeably different modules in terms of content.

The design of specialized educational programs for the formation of both professional and information-mathematical competencies should be at an advanced level exceeding just the simple modernization of existing educational programs. It is assumed that the new programs are of the maximum possible variable nature, allowing the construction of individual educational learning trajectories, taking into account the development of the system of social values of students in the digital educational environment. A characteristic feature of such programs is the rational combination of the principles and technologies of traditional and digital didactics, preserving their distinctive advantages. The model for designing the content of one of the training programs is shown in Fig. 1.

The criteria for selecting the content of training are both general didactic principles (consistency; professional orientation; focus on the individual; humanization) and particular methodological principles (relevance and contextuality; historicism; professionally significant social values of the individual). The use of specialized educational programs for the preparation of students in the conditions of self-isolation of participants in the educational process showed the presence of both positive and

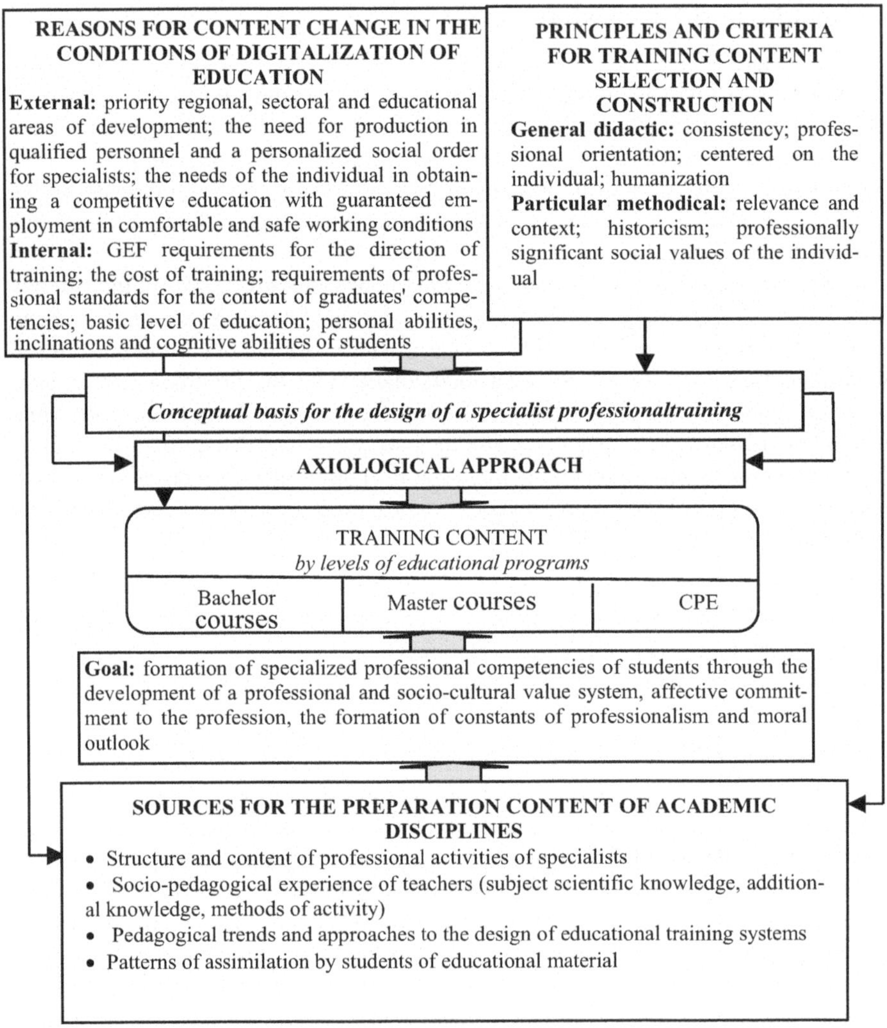

Fig. 1 Model for designing the content of professional training in the context of digitalization of education

negative factors affecting the educational process, depending on changes in learning conditions. The positive factors were:

- Formation of additional digital competencies among students when working with new electronic platforms and services;
- Availability of social freedom and mobility of an educational process participants;
- Integrating of new digital didactics and new scientific knowledge in the field of pedagogy.

In principle, these results confirm the validity of the above ideas about the need to change the methodology of teaching academic disciplines towards their possible integration.

The factors negatively affecting the quality of the educational process during the period of limited opportunities for contact work with students include:

- Decrease in educational and executive discipline of students, educational and cognitive interest;
- Emergence of communication disorders among young people, reducing the effectiveness of their socialization;
- Emergence of a constant stress syndrome and a decrease in the empathy level among all educational process participants.

The statistics of educational achievements of TSTU students over the past three years (average grade) suggests that the transition to improving educational programs has minimized the negative consequences of the impact of the pandemic while limiting contact work with students. In our assessment, we adhered to the recommendations of [12]. The negative aspects of distance learning identified in the course of the study aim at further improvement of educational programs and, first of all, at the development of the psychological component. [13]

4 Conclusion

The organization of training during the pandemic clearly confirmed the thesis that the era of the digital personality and new digital scientific and pedagogical knowledge has come, which ensures greater stability in the development of mankind.

Distance learning as a method of overcoming the difficulties of self-isolation of participants in the educational process is a fait accompli that requires the justification of new didactic approaches and the development of a system of social values of students in the digital educational environment. The results obtained make a scientific contribution to the pedagogical process of designing educational programs in conditions of limited contact work with students (the period of the COVID-19 pandemic), as well as to the methodology for transforming students' personality and value system, in order to preserve the health and emotional intelligence of all participants in the educational process. It is required to continue the interdisciplinary study of various aspects of the digitalization of education, taking into account the characteristics of the studied academic disciplines.

In the context of limited opportunities for contact work of teachers and trainees, the implementation of specialized educational programs in mathematics that rationally combine the advantages of analytical methods of classical mathematics and advanced digital information technologies, principles and technologies of traditional and digital didactics reduces the level of disadvantages of distance learning.

In conditions of mutual self-isolation, it is necessary to shift the values' paradigm of the educational process participants on a focus of social orientation, and prompt elimination of factors that negatively affect the educational process.

A new approach to the design of educational programs, on the example of the Tambov State Technical University, has confirmed its continuity for use in the higher education system.

Acknowledgements This work was supported by the Russian Foundation for Basic Research (project no. 20-08-00091).

References

1. Rahiem MD (2020) Technological barriers and challenges in the use of ICT during the COVID-19 emergency remote learning. Univ J Ed Res 8:6124–6133
2. Almaiah MA, Al-Khasawneh A, Althunibat A (2020) Exploring the critical challenges and factors influencing the E-learning system usage during COVID-19 pandemic. Ed Inf Technol 25:5261–5280
3. Apriyanti C (2020) Distance learning and obstacles during Covid-19 outbreak. J Ilmiah Pendidikan Dasar 7(2):68–83
4. Dorohova TY, Puchkov NP (2021) Development of individual training programs at a technical university. Quest Mod Sci Practe Univ VI Vernadsky 1(79):137–146
5. Buckingham D (2007) Media education goes digital: an introduction. Ed Media Technol 32(2):111–119. https://doi.org/10.1080/17439880701343006
6. Mailizar, Almanthari, A, Maulina S, Bruce S (2020) Secondary school mathematics teachers' views on e-learning implementation barriers during the Covid-19 pandemic: the case of Indonesia. Eurasia J Math Sci Technol Ed 16(7):186–190
7. Droessiger G, Vdovinskene S (2020) Motivation factors for attending theoretical classes among students of technological specialties. Integr Ed 24(1):50–61. https://doi.org/10.15507/1991-9468.098.024.202001.050-061
8. Bacolod DB (2022) mobile learning as a solution for restricted learning during the COVID-19 pandemic. J Digit Ed Technol 2:ep2203. https://doi.org/10.21601/jdet/11584
9. Al Majali S, Al Kaaldi K (2020) Values of tolerance on academic achievement, cultures and gender among UAE university students. Int Ed J 13(3):571–586. https://doi.org/10.29333/iji.2020.13339a
10. Xue E, Li J, Li T, Shang W (2020) China's education response to COVID-19: a perspective of policy analysis. Ed Philos and Theory 5:1–13
11. Lobanova NI, Puchkov NP (2021) Digitalization of mathematical education: teaching the course "differential equations". Iss of Modern Sci and Pract Univ VI Vernadsky 2(80):128–139

12. Jacques S, Ouahabi A, Lequeu T (2020) Remote knowledge acquisition and assessment during the COVID-19 Pandemic. Int J Eng Pedag 10:120–138. https://doi.org/10.3991/ijep.v10i6.16205
13. Dorohova TY, Puchkov NP (2021) Information-mathematical competencies of students in the system of targeted training of engineering personnel. Continuum 2(22):16–22

Application of Mathematics History in Integral Calculus Teaching Under Conditions of Social Isolation

Inessa V. Ignatushina and Inna K. Zubova

Abstract The aim of the research is to develop an algorithm of including mathematics history in teaching students who study mathematical disciplines. To achieve this goal, it is necessary to solve the following tasks: the analysis of the most successful practices of using historical digressions in teaching students; the development on their basis of the algorithm synthesis for including elements of mathematics history in teaching mathematical disciplines, in particular, mathematical analysis; approbation of the developed algorithm in educational process of technical and pedagogical universities. The article considers one of the approaches to historical and mathematical information inclusion in teaching of such an important section of mathematical analysis as integral calculus. Didactic and educational aspects of using information on history of integral calculus formation in teaching bachelors under conditions of social isolation caused by COVID-19 are shown.

Keywords Calculus teaching · Social isolation · Integral calculus · Distance learning · History of mathematics

1 Introduction

Mathematical analysis is one of the most difficult and most important disciplines in the curriculum of any mathematical or technical specialty. A first-year student often encounters a completely new, unusual approach to even the most basic concepts of this discipline. This issue is exacerbated by the social isolation caused by COVID-19. Therefore, when lecturing, it is especially important to constantly emphasize that there is nothing far-fetched in mathematics, that life itself, in the course of the development of this science, posed issues for people, the solutions of which

I. V. Ignatushina (✉)
Orenburg State Pedagogical University, 19, Sovetskaya St, Orenburg 460014, Russia
e-mail: streleec@yandex.ru

I. K. Zubova
Orenburg State University, 13, Pobeda Ave, Orenburg 460018, Russia

A. L. Semenov et al. (eds.), *Structural and Technological Transformation of Education in the Post-Pandemic Period*, Lecture Notes in Networks and Systems 1282,
https://doi.org/10.1007/978-3-031-84039-5_12

inevitably led to the introduction of new mathematical concepts and the creation of rigorous scientific theory. Historical analysis of a particular issue appearance often helps to answer the question of its origin, the significance of its solution. This, in its turn, facilitates the perception of some relevant educational material and enables to develop a more general view on the issue under consideration. Therefore, historical and mathematical awareness should be an integral element of the formation of the professional culture of a modern specialist [1].

2 Survey Methodology

The study was carried out in the following sequence: from the study of theoretical issues of mathematics history application in teaching integral calculus and the analysis of the advanced pedagogical world experience in this field to the development of tools for the implementation of this technology at the university. The FSEI HE "Orenburg State Pedagogical University" and the FSEI HE "Orenburg State University" became the basis for the experimental work.

2.1 Related Work

It was written about the benefits of using historical excursus in teaching different branches of mathematics by the first Russian historian of mathematics Bobynin V. V. [1] and well-known methodologists D. Galante [2], K. Clark [3], P. Liu [4], W. T. Jankvist [5] and others. Nor should we fail to note here such authors of fundamental works and textbooks on the history of mathematics and mathematical education as K. A. Rybnikov [6], T. S. Polyakova [7], M. F. Gilmullin [8], Yu. A. Drobyshev [9]. In Russia the issues of theory and practice of teaching mathematical analysis at university were developed by Zhuravleva N. A. [10], Zadorozhnaya O. V. [11], Kalinin S. I. [12], Smirnov E. I. [13] and others. However, the issue of historical and scientific information application in the process of teaching integral calculus has not been studied enough.

3 Results and Discussion

In the context of this work, under historical and scientific information we mean the information about the following historical and scientific facts: 1. The moment when the task, the solution of which required the development of new concepts and methods, was set; 2. The moment when such task was fully or partially solved; 3. The publication of a scientific work or the appearance of the document (letter, entry in a diary or notebook, etc.), which presents a formulation, an attempt to solve or a complete solution of such issue.

It is such kind of information that allows carrying out a historical analysis of the solution of the issues important for the further development of mathematics and forming basic concepts of mathematical analysis in the process of their solution.

3.1 Survey Findings

Let us consider the algorithm for including mathematics history in teaching integral calculus (Fig. 1).

To detail this algorithm, let us analyze specific examples of historical and scientific information use as a means of assisting students in integral calculus study.

It is the most convenient to introduce students to the integral concept formation history after the presentation of the topics "Antiderivative function", "Indefinite integral" and "Basic methods of integration". Next, we introduce a concept of definite integral. It is usually introduced "according to Riemann", i.e., is defined as the limit of integral sums. After a brief biographical reference describing the role of Bernhard Riemann (1826–1866) in the development of modern mathematics, it should be noted that the formation of the concept in question was long and difficult. The

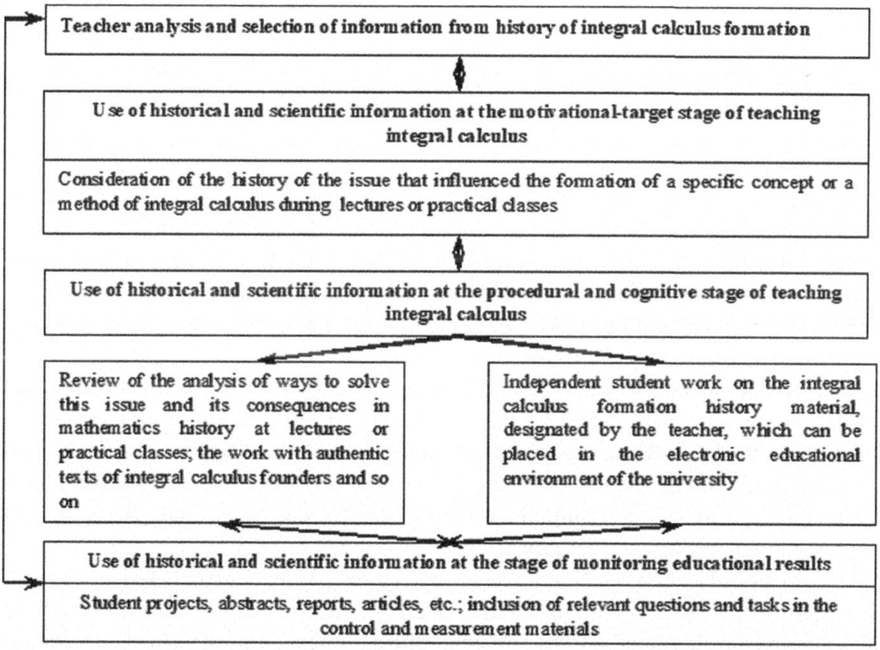

Fig. 1 The algorithm of inclusion of mathematics history in teaching integral calculus

idea that started that journey, originated in the works of Archimedes (c. 287–212 BC) at the search for areas of curvilinear figures and volumes of some bodies of revolution. The results of the scientist's research were presented in the geometric form usual for Greek mathematics. Finding, for example, the area of a parabolic segment, Archimedes divided the height of the figure into a large, but finite number of segments of very small but finite length. These segments became the heights of the rectangles, from which the stepped figures contained in the parabolic segment and containing it were composed. Speaking about the geometric meaning of integral sums, it should be noted that these stepped figures are the prototype of the concepts of upper and lower integral sums.

Archimedes' method was developed in the works of Johannes Kepler (1572–1630). With its help, Kepler checked the results of Archimedes, found the areas of new flat figures and the volumes of new bodies of revolution. Trying to unravel the idea of Archimedes, he represented any figure (or body) as a sum of many small elements, then from these elements (deformed, if necessary) he composed a new figure (body) with an already known area (volume). In this way, he obtained the required area of a given figure or the volume of a given body.

Many scientists of the 16th–seventeenth centuries devoted their work to the enhancement of Kepler's method. The most famous is the theory of "indivisibles", which was developed by Bonaventure Cavalieri (1597–1644). He introduced the term "indivisible", thereby approaching the concept of an infinitely small element.

Comparing "indivisibles", i.e., small elements of various figures, among themselves, and considering that, for example, the areas of these figures are in the same relation to each other as their "indivisibles", Cavalieri, knowing the areas of some figures, found the areas of others. Using this method, Cavalieri actually calculated integrals of the form.

$\int_0^a x^n dx = \frac{a^{n+1}}{n+1}$,

where n is a positive number.

It is easy to see the similarity of the considered methods with modern integral methods of solving geometric issues. However, although the idea of limit was present in them, still there was no strict definition of this concept. It is very important to analyze in detail the similarities and differences between the methods of Archimedes, Kepler and Cavalieri with modern integral methods. This will contribute to a deeper understanding of the essence of the "definite integral" concept.

Several new results in the calculation of areas and volumes in the seventeenth century was received by Pierre Fermat (1602–1675). Introducing the division of a squarable figure by ordinates spaced from each other at equal distances, he developed a new method for calculating the areas of curvilinear trapezoids enclosed under a curve of the form $y = x^n$, suitable for both integer and fractional, positive, and negative values of the index of power except for the case $n = -1$.

The results obtained by Newton's teacher Isaac Barrow (1630–1677) were of paramount importance for the nascent analysis of infinitesimal. He established the mutually inverse character of issues on quadratures and tangents in terms of geometry, i.e., from the point of view of modern mathematical analysis, the mutually inverse nature of issues of integral and differential calculus. It is appropriate to mention this when speaking about the geometric meaning of such concepts as the derivative and antiderivative.

Moving on to calculating definite integrals, students become familiar with the Newton-Leibniz formula. Here it is worth dwelling on how the founders of differential and integral calculus interpreted the concept of a definite integral.

In the second half of the seventeenth century Isaac Newton (1643–1727) and Gottfried Wilhelm Leibniz (1646–1716) independently created a new calculus—the apparatus of the emerging mathematics of variables. Newton presented it in the form of the "theory of fluxions" (as he and his followers called derivatives), and Leibniz—in the form of "calculus of differentials." The fact that in both cases we can talk about the creation of calculus is demonstrated by the evidence that both great mathematicians, each in his own way, defined the basic concepts of mathematics of variable quantities, established their correlation, introduced suitable symbols and developed a calculation algorithm. It is appropriate to explain to students that in modern mathematics the term "calculus" means an apparatus used in some area of mathematics. The calculus is formed when the basic concepts defining a certain set of mathematical models, the rules of procedure with these models, terminology and symbols are introduced.

When integrating, I. Newton used a mutually inverse connection between finding fluxions (derivatives) and fluents (functions). In this regard the representation of integrable function by a power series played an important role.

G. V. Leibniz interpreted a definite integral as the sum of an infinite number of infinitesimal elements, which he called differentials. He originally designated that integral as "omnl" (omnia is all, l is a line, that is, "all lines"), but soon he introduced the sign $\int$, which was an elongated letter S stretched up beyond recognition, the first letter of the word summa—"sum". The term "integral" itself was put into practice by one of the first disciples of Leibniz—Johann Bernoulli (1667–1748).

Leibniz, like Newton reduced the calculation of integrals to finding antiderivative. A constant of integration appeared in an article by Leibniz in 1694. To calculate integrals with certain limits of integration, Newton and Leibniz, each with the help of his own symbolism, used the formula that now bears their names. The Newton-Leibniz integral is equal to the difference between the values of the antiderivative at the points that are the upper and lower limits of integration.

Leonard Euler (1707–1783) made an enormous contribution to the development of integral calculus. Almost all the methods of integration described in modern textbooks can be found in his three-volume "Integral Calculus". In addition, Euler significantly developed the theory of definite integrals, for which he introduced a special sign $\int Pdx\left[\begin{smallmatrix}\text{from } x=a\\ \text{to } x=b\end{smallmatrix}\right]$. The term "definite integral" itself was offered by Pierre Simon Laplace (1749–1827) in 1779 and the modern notation for a definite integral $\int_a^b f(x)dx$ was introduced by Jean-Baptiste Fourier (1768–1830) in 1817.

In conclusion of this excursus, it should be emphasized that the introduction in the nineteenth century of a strict definition of a limit made it possible to give a modern formulation of a definite integral. Here we can once again mention the activities of Augustin Louis Cauchy (1789–1857), pointing out that in the modern course of mathematical analysis the concept of a limit is introduced precisely "according to Cauchy", and all other important concepts of the course including the concept of a definite integral are based on this very definition.

3.2 Data Analysis

Experimental work on the introduction of historical excursus in students teaching as a means of assisting them in integral calculus study was carried out on the basis of the Physics and Mathematics Faculty of the FSEI HE "Orenburg State Pedagogical University" and the Faculty of Mathematics and Information Technologies of the FSEI HE "Orenburg State University". It was carried out in three stages: 1. Search and ascertaining stage of the experiment (2004–2010); 2. Formative stage of the experiment (2010–2020); 3. Generalizing stage of the experiment (2020–2021).

At the first stage, the corresponding psychological, pedagogical and methodological literature was being studied. The curricula, programs and electronic educational environment of universities on mathematical analysis and integral calculus, as its section were being analyzed. Conversations on an issue of interest with students and teachers were being conducted. The work of students was being observed during lectures and practical classes on integral calculus. A questionnaire survey and students assessment analysis, including a series of tests and mathematical dictations on the main sections of integral calculus, were being carried out. Possibilities of using mathematics history in teaching students as a means of assisting them in integral calculus study were being determined. The algorithm for historical and scientific information inclusion in the study of integral calculus was being developed.

Students from 14 academic groups with a total number of 242 people took part in this stage of the experiment. The results of the tests and mathematical dictations on the main sections of integral calculus performed by them made it possible to determine the educational material assimilation coefficient (EMAC) [14] in each group and the coefficient of knowledge consistency (CKC) [15].

The results are shown in Table 1.

Table 1 Results of search and ascertaining stage of the experiment

Year	Number of respondents		EMAC		CKC	
2004	22	21	0.63	0.61	0.55	0.52
2005	21	20	0.60	0.62	0.51	0.52
2006	20	22	0.60	0.59	0.53	0.53
2007	17	19	0.60	0.58	0.54	0.55
2008	20	20	0.61	0.60	0.55	0.53
2009	21	23	0.63	0.63	0.55	0.54
2010	19	20	0.65	0.62	0.56	0.53

Table 2 Results of the formative stage of the experiment

Year	Number of respondents		EMAC		CKC	
2011	20	21	0.66	0.65	0.59	0.57
2012	21	22	0.64	0.72	0.58	0.67
2013	18	21	0.64	0.69	0.61	0.69
2014	21	23	0.68	0.63	0.66	0.56
2015	20	20	0.69	0.72	0.68	0.69
2016	23	24	0.67	0.6	0.70	0.71
2017	18	21	0.73	0.74	0.71	0.72
2018	21	22	0.72	0.75	0.70	0.76
2019	17	20	0.81	0.82	0.80	0.81
2020	20	19	0.80	0.78	0.79	0.80

Based on these results, we can conclude that the students' knowledge of integral calculus is formal, the respondents have little idea of the relationships between its main sections, as well as the areas of its application. In addition, most of them have a positive attitude toward studying the material on integral calculus history.

At the second, formative stage of the experiment, the developed algorithm of historical and scientific information inclusion in the study of integral calculus was tested.

As a result, students became familiar with the reasons for the appearance of integral calculus, the mathematical laboratory of its founders, as well as its applications, what helped them to determine the place of this section of mathematical analysis in the science as a whole, look at it on a larger scale, and realize its great practical significance.

The participants of this stage of the experiment (412 students from 20 academic groups) were also surveyed. They were asked to perform the same series of tests and mathematical dictations as the participants at the first stage of the experiment. The obtained coefficients of the educational material assimilation and the consistency of knowledge are presented in Table 2.

4 Discussion

The effectiveness of the developed algorithm for historical and scientific information inclusion in integral calculus study was tested at the third, generalizing stage of the experiment. The Mann–Whitney U-test was used to compare the results of the control group shown in Table 1 with the results of the experimental group presented in Table 2.

Since the calculated values of the criterion ($U_{\text{empfor EMAC}} = 7$, $U_{\text{emp for CKC}} = 2$) turned out to be significantly less than the critical value at the significance level $\alpha = 0.05$ ($U_{0.05} = 92$), the observed difference in the values of the educational material assimilation coefficient (coefficient of knowledge consistency) statistically significant for the control and experimental groups. At the same time, the results of the experimental group are higher. Consequently, the use of historical excursuses in teaching students integral calculus has a positive effect on the increase of coefficients of educational material assimilation and the consistency of knowledge in this section of mathematical analysis.

5 Conclusion

Thus, the use of historical and scientific material in lectures, practical classes and in the electronic educational environment of the university helps students to link together all the information they know about the basic concepts of integral calculus. Students' ideas about these concepts become clearer. The results of the experiment show that, using historical and scientific facts, teacher demonstrates the development path of the studied mathematics section to students. This makes it easier for students to understand the importance of this section for the whole mathematical science. In addition, this enables to form moral ideals based on personalistic component of mathematics history among students. And all of this contributes to a more conscious and successful mastering of educational material. For educators, "historical background provides insights and information on which we can base our approaches to teaching" [6]. Under conditions of social isolation caused by COVID-19, the use of mathematics history in teaching integral calculus online or in a mixed format gave very good results. This conclusion can be reached if we look at the coefficients of educational material assimilation and consistency of knowledge for 2019 and 2020, which are reflected in Table 2. They turned out to be higher than in previous years. Therefore, the use of the historical component in teaching integral calculus is effective in different formats of lessons: offline, online and mixed.

References

1. Bobynin VV (1886) Philosophical, scientific and pedagogical significance of the history of mathematics. Physics and mathematics in their present and past, Moscow
2. Galante D (2014) The use of the history of mathematics in the teaching pre-service mathematics teachers. Redimat 3(2):110–120
3. Clark K, Kjeldsen T, Schorcht S, Tzanakis C, Wang X (2016) History of mathematics in mathematics education. Recent developments. In: History and pedagogy of mathematics. hal-01349230. https://hal.archives-ouvertes.fr/hal-01349230
4. Liu P (2003) Do teachers need to incorporate the history of mathematics in their teaching? Math Teach 96(6):416–421
5. Jankvist UT (2010) An empirical study of using history as a 'goal.' Ed Stud in Math 74(1):53–74. https://doi.org/10.1007/s10649-009-9227-8
6. Rybnikov KA (1981) On the historical and methodological foundations of mathematical education of teachers. Math Sch 5:31–33

7. Polyakova TS (1998) Historical and methodological training of mathematics teachers at a pedagogical university. Dissertation, St. Petersburg
8. Gilmullin MF (2009) Formation of the historical component of the mathematical and methodological culture of students in teaching the history of mathematics at a pedagogical university. Dissertation, Yaroslavl
9. Drobyshev YA (2011) Multilevel historical and mathematical training of the future teacher of mathematics. Dissertation, Moscow
10. Zhuravleva NA (2012) Formation of basic key competencies of students—future teachers of mathematics—in the process of teaching mathematical analysis. Dissertation, Krasnoyarsk
11. Zadorozhnaya OV (2011) Designing a complex of educational projects in the process of teaching mathematical analysis at the university. Dissertation, Nizhny Novgorod
12. Kalinin SI (2010) Methodical system of teaching pedagogical university students differential and integral calculus of functions in the context of fundamentalization of education. Dissertation, Moscow
13. Smirnov EI, Shabalina AI (2009) The principle of variability in the design of the spirals of foundation of knowledge in mathematical analysis. Paper presented at the 7th international kolmogorov readings. Yaroslavl State University Publishing, Yaroslavl
14. Bespalko VP (1989) Components of pedagogical technology. Pedagogika, Moscow
15. Kuznetsova NE (1989) Formation of a system of concepts in teaching chemistry. Prosveshchenie Publishing, Moscow

Information Technologies in Blended Learning of Mathematics at University in the Context of the COVID-19 Pandemic

Natalia N. Yaremko, Vladimir Selutin, and Margarita Gavrilova

Abstract These days, due to the circumstances caused by the Covid-19 pandemic, universities had no other choice but to urgently transfer the process of learning to the digital form. The role of information technologies (Information Technologies – IT) in the current learning environment becomes predominant. The blended form of learning (Blended Learning – BL) becomes the main one under quarantine measures as the most effective. The purpose of this research was to determine the current state of the global digitalization of universities, as well as to analyze, generalize experience, and make a forecast for it in the context of the COVID-19 pandemic in terms of the use of information technologies in the blended learning of mathematics at a university. In this research we defined periods of development of blended learning, identified three predominant models of blended learning at a university, analysed the use of information technologies in blended learning of mathematics at a university. We dedicated invariant component of a mathematical content in the blended form of learning. In this paper we made a diagnostic of satisfaction of students and lecturers with the blended form of learning with the use of information technologies in the teaching of mathematics and built a forecast of the further development of blended learning of mathematics at a university in the post-pandemic period. Three universities of the Russian Federation served as a research base: NUST MISIS, Moscow, Penza and Orel State Universities; 322 students and 56 lecturers took part in the survey for three years (2019–2022).

Keywords Information technologies · Blended learning · Teaching and learning mathematics · Visualization

---

N. N. Yaremko (✉)
NUST MISIS, 4, Leninsky Prospect, 119049 Moscow, Russia
e-mail: yaremki@yandex.ru

V. Selutin
Orel State University, 95, Komsomolskaya St, 302026 Orel, Russia

M. Gavrilova
Penza State University, 40, Krasnaya St, 440026 Penza, Russia

A. L. Semenov et al. (eds.), *Structural and Technological Transformation of Education in the Post-Pandemic Period*, Lecture Notes in Networks and Systems 1282,
https://doi.org/10.1007/978-3-031-84039-5_13

1 Introduction

The special significance of mathematics in the twenty-first century is reasonably recorded in the report made by the Mathematical Association of America (MAA) "The Importance of Mathematical Sciences at Colleges and Universities in the 21st Century. Importance to Society, to Students, to K-12 Education, to Departments", The leading role of mathematical methods and models in the development of modern technologies is recognized in many Russian and foreign studies [1–8]. At the same time, our experience of teaching mathematics to students using academic approaches indicates significant difficulties in the study and mastering of its content by modern students. The authors have been searching for a solution to the problem of accessible dialogue with students in the study of mathematics for many years. The study of foreign researches in this area has made it possible to highlight the use of blended learning as a priority [9–12].

Blended Learning first appeared as a form of education in the sixties of the XX century in corporate and higher education. The term itself – Blended Learning (BL) – dates back to 1999, when the American Interactive Learning Center began releasing software designed for teaching over the Internet. The most general definition of blended learning is given in the works of [13–15] "Blended learning (BL) involves integrating in-person and online instruction" The definition of BL in terms of student activity is given in [16] BL is "at least in part through online learning, with some element of student control over time, place, path and/or place". Describing blended learning, [14] say that blended learning represents a significant departure from any of the approaches. It represents a fundamental rethinking and reorganization of the dynamics of teaching and learning, starting with different contextual needs and unforeseen circumstances (for example, discipline, level of development, and resources) Based on this, there are no two identical models and definitions of blended learning. This is what makes the whole complexity of this approach.

The period after 2006 can be called a period of collection of experience, accumulation, time of practical implementation of BL ideas. This is the time of active use of blended learning: there is a development of terminology, methodology of blende learning, application of the format of blended learning in various educational institutions. We come to understanding of the advantages and awareness of the disappointments in the first mass applications of blended learning, mistakes, and their analysis. [1, 2, 7, 9]. The use of blended learning in Russia is hindered by the low availability of technical equipment in educational institutions, the high cost of equipment (computers, tablets, lack of developed content), the lack of methodological course books and the unpreparedness of lecturers for such a teaching format. Breakthrough in the use of blended learning and information technologies dates back to 2019. The pandemic acts as a catalyst for accelerating the introduction of information technologies and blended learning (BL) into the practice of universities. The blended form of education is becoming predominant, [5, 9, 10, 12, 18]. By this time, the conditions for the introduction of BL were fine: information technologies have been developed, there is an information and educational environment in universities,

LMS have been introduced, there is a stable Internet connection, a sufficient number of digital educational resources and tools [19].

The purpose of this study is to determine the current state, as well as to analyze, generalize experience, build a forecast of the use of information technologies in the blended learning of mathematics at university in the context of a pandemic on the example of universities in three cities of Russia: Penza, Orel and Moscow. The article solves the following tasks:

- the stages of the formation of blended learning have been identified and characterized;
- the application of blended learning technologies (Self-Blend, Rotation, Flipped ClassRoom, Flex), IT tools, digital and electronic resources have been analyzed;
- the invariant component of the mathematical content in the blended form of learning has been determined;
- diagnostics of satisfaction of students and lecturers with a blended form of learning with the use of information technologies in teaching mathematics has been carried out;
- a forecast of the further development of blended learning with the use of information technologies in teaching mathematics at university in the post-pandemic period has been built.

Three universities of the Russian Federation served as a research base: NUST MISIS University, Moscow, Penza and Orel State Universities; 322 students and 56 lecturers took part in the survey for three years (2019–2022).

2 Survey Methodology

In our research, we were guided by normative and program documents that regulate the main goals and requirements for the preparation of students at university conclusions of foreign and Russian scientists on the organization of blended learning [13, 16], the use of information technologies in education [9, 10, 12, 18], as well as the results of surveys of lecturers and students.

The main research methods involved theoretical analysis and comparison of domestic and foreign approaches to the study of mathematics in higher education in the context of the COVID-19 pandemic. To solve the tasks set, a set of theoretical, practical-experimental, and empirical research methods was used.

The basis of practical and experimental research methods were the criteria for identifying the stages of research, the logic of direct classroom and distance learning with students, as well as the purposeful search for forms of blended mathematics learning in a particular university, the expediency of restructuring the content of mathematical disciplines. Observation, questioning, testing, and polling are the main components of empirical methods at each of the three stages of the research. Our research was aimed at analyzing, evaluating, and forecasting the organization of blended mathematics learning in context of pandemic and studying the impact of information technologies on this process.

3 Results and Discussion

The first results of the experimental work made it possible to identify the stages of the formation of blended learning, to determine its advantages and disadvantages. It was found out which main software tools were used by universities at each stage and which of these tools make it possible to ensure the effectiveness of blended learning. A systematic analysis of personal experience and polling results allowed us to build a model of blended learning in each of the universities, to choose a single multifunctional platform for the organization of learning for the university (LMS CANVAS is in Moscow, LMS MOODLE is in Orel and Penza). The methodological basis of the research, principles were identified, and the conceptual apparatus was formed. A description of general approaches to structuring mathematical material and organizing classes in the implementation of blended learning using the LMS CANVAS and LMS MOODLE platforms was given.

The content of the educational course hosted on the LMS CANVAS (Moscow), LMS MOODLE (Orel, Penza) platform highlights separate topics – modules, there are usually 3–4 of them per semester. Each of the modules includes theoretical material, presentations, examples of solving problems, introduction to video clips of experiments, laboratory works, training tasks and check-up. Intermediate check-up of the module material mastering is carried out in the form of two control measures: a classroom check-up and an individual homework. The classroom check-up can be carried out in the form of a test. The students "attach" the completed classroom test and individual homework in the appropriate section of the course. Students can view their verified papers with ratings and comments in the "Comments" section. On the LMS CANVAS and LMS MOODLE platforms, it is possible to organize student–lecturer and student–student interaction using the forum functions.

Thus, the course on the LMS Canvas platform in mathematics in accordance with the program of study at NUST MISIS in the 2nd year in the third fall semester of the 2021–2022 academic year consisted of 3 modules: Module 1. Series; Module 2. Theory of functions of a complex variable; Module 3. Operational calculus.

Since the beginning of the fall semester of the 2021–2022 academic year at NUST MISIS, mathematics learning has been organized in a blended format, since some students were quarantined due to the epidemic and attended classes remotely, while the other part was present in the classroom. Moreover, after some time, the composition changed: students who attended classes remotely, switched to full-time study; full-time attendees, on the contrary, switched to distance learning. In such conditions - the constant change of the attendees' proportion – the blended format of learning was almost the only way out of the circumstances.

In our opinion, in the conditions of blended learning, the use of "Flipped Classroom" technology has allowed to brisk up the educational process, stimulate the work of students both in the classroom and those who study online. [20, 21] Before conducting a practical lesson, students were asked to study a video fragment as a homework assignment, make notes and answer test questions. Then, in the next class, start solving tasks in small groups – discussion rooms in the LMS Canvas or in the classroom for students who were present in person. Upon completion of the decisions – discussion, conclusions, evaluation of the work. This technique was used in the study of the topics "Calculation of residues" and "Application of residues." All the necessary materials were attached to the course in the LMS Canvas. Feedback was provided via chat or using the voice capabilities of MS Teams.

As an experiment, IT tools for solving mathematical problems were tested. These are online free-to-use calculators, licensed and free versions of WolframAlpha, MatLab, MathCad, Excel. A high-quality educational IT product is presented by 1C company "1C Mathematical Constructor. Virtual laboratories, simulation modeling". During the pandemic, it is freely available.

Thus, to implement blended learning with the use of information technologies, the content of education was revised, educational content was developed in accordance with new accents and interpretations, and digital educational tools were selected.

General results. The experience of learning mathematics in a pandemic (2019–2021) shows that there is a gradual replacement of the academic system of learning with a blended one, with active use EIOS of universities (Electronic information educational environment), LMS platforms, video lectures and IT tools designed to solve mathematical problems.

The functionality of various LMS platforms is sufficient to organize a blended learning of mathematics at university. Interesting online courses in mathematics are presented on the platforms like Stepik, Coursera and OpenEducation. Recently, many of the world's leading universities have made their video lectures freely available. You can use lectures on mathematical disciplines from MIT (USA), Moscow State University (Russia), PhysTech/MIPT (Russia), Tomsk Polytechnic University (Russia).

The leading role of mathematical methods in the development of all spheres of society is emphasized in the report of the Rector of Moscow State University [10] at the All-Russian Congress of teachers and lecturers of Mathematics and Computer Science. According to many authors [13, 22], mathematical education allows the effective formation of digital competencies. [23–25]

We consider the stage of organizing check-up procedures to be the most difficult and ambiguous: the lack of independence in completing tasks, the use of outside help by students, the large time spent by lecturers on creating contextual tasks, finding out the originality and independence of the completed control task are negative factors. Automatic generation of tasks and their completion statistics are positive factors. The factors that both contribute to and hinder effective blended learning, as the main one in the current situation of a returning pandemic, are identified.

The generalized results of the study are presented in Fig. 1. Intensive growth is noticeable in all the main directions of the organization of the process of learning mathematics. Selective research on the preference of the learning format was

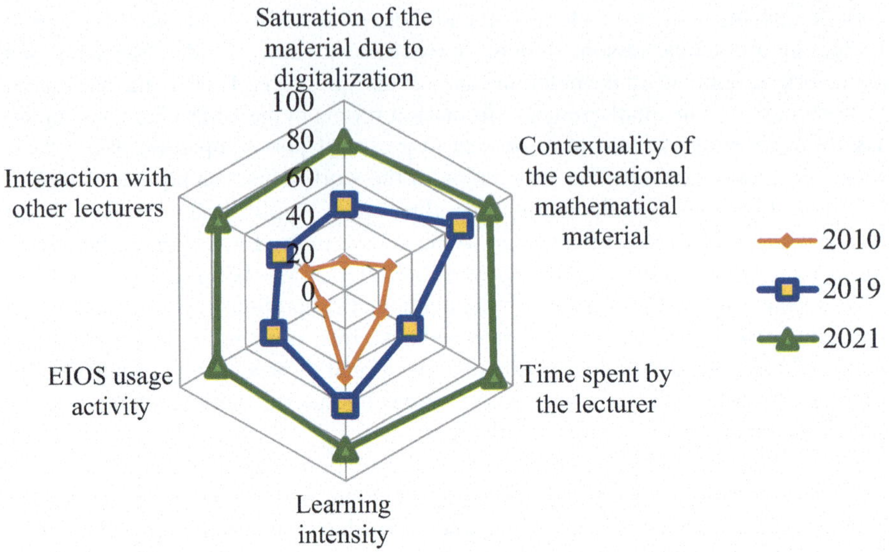

Fig. 1 The influence of blended learning in mathematics on the main factors of the organization of the educational process

conducted at the initial stage of the pandemic and during the formation of blended mathematics learning at universities.

At the initial stage of the pandemic, the majority of lecturers (76%) preferred full-time education. Students (72%) selected distance learning. Over the past time, we have seen a convergence of the positions of lecturers and students when choosing a learning format.

The restrained attitude of lecturers to blended learning of mathematics (46%) is due to many factors, the main of which are: the need to acquire and constantly train to work with a variety of equipment and software and information technologies; lack of development of methods for using various tools for teaching mathematics, unpreparedness of technological section of the educational process. Students prefer a blended learning format (61%), as it combines elements of the freedom of distance learning and allows them to discuss and solve problems together during the period of full-time education. Thus, if we follow the general prevailing opinion of the educational community, then blended learning should become the leading form of education not only during the pandemic, but also in the post-pandemic time. At the same time, satisfaction with the process of learning mathematics from both students and lecturers does not exceed 60%. The main problems of implementing information technologies are presented in Fig. 2.

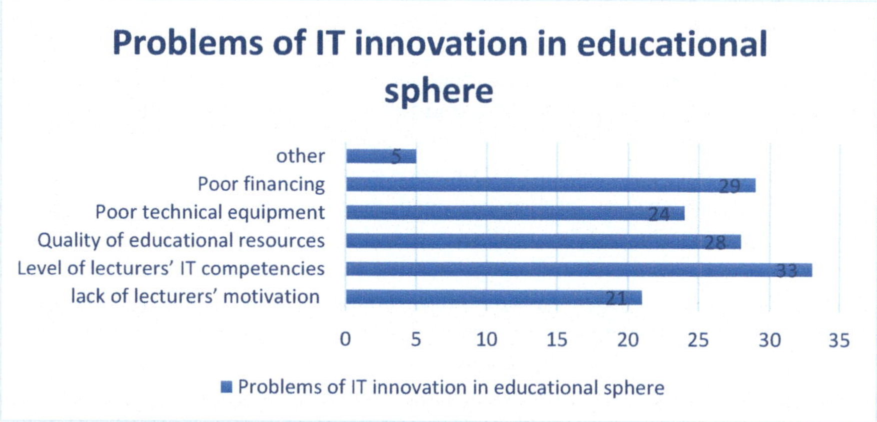

Fig. 2 A ranked list of the main problems associated with the introduction of information technologies in the process of learning mathematics

4 Conclusion

Studying the process of learning mathematics at university students in the conditions of the COVID-19 pandemic allowed us to draw the following conclusions: blended learning is a large-scale strategy that helps break down the stereotypes inherent in the traditional learning system. It gives lecturers and students the opportunity for broad cooperation and communication that were previously impossible. This is manifested in the change of organizational forms and methods: the presentation of new material is the technology of the "inverted class", the allocation of the leading idea, information saturation based on the use of various information sources, including VR/AR technologies, flexible schedule, working in small groups, mutual check-up, self-check. The use of various LMS platforms, modeling software, and information technologies contribute to changing the content of mathematical courses in the context of their modularity, visualization, creation of mathematical models, reduction of routine calculations, orientation to the problematic nature of information and the formation of critical understanding of information. There is a simultaneous development of mathematical and digital competence.

Disadvantages. The variety of IT tools makes it difficult for lecturers to choose them, the lack of methodological materials requires constant independent mastering of new software products by lecturers, the lack of high-quality multimedia materials for each section of mathematics gives rise to unsystematic use of these materials, overload of classes with demonstrations and visual images to the detriment of the development of logic, evidence, critical judgments.

The prospects. What kind of mathematics to teach in the twenty-first century? It is necessary to preserve the fundamental nature, logic, and evidence-based judgments. Perform routine, technically complex computing work with the help of IT tools. Teaching mathematical modeling and its methodology should become predominant;

the formation of critical and systematic thinking of students should become a priority goal in teaching mathematics. The necessary additional resource is the development of the digital environment of the university, as well as the development of multilateral cooperation, the enrichment of the traditional content of disciplines with video lectures, VR situations, and the use of a set of contextual tasks.

Augmented (AR) and virtual (VR) reality seem to be promising technologies for organizing the process of learning mathematics. They can be attributed to the most important innovations that are advisable to be used in the educational process. In education, the effectiveness of these information technologies is based on several factors: visibility, visualization, dynamism, changeability of images, cognitive interest of students, emotional coloring of learning, the degree of student involvement, the mastering of the material.

The results of the research provide an assessment of the current state and trends in the development of the process of learning mathematics at university – the development of various forms of blended learning based on the digitalization of this process. Learning mathematics in the twenty-first century will not be able to remain the same [6, 26], the world of education and training is undergoing significant changes now and will continue to undergo significant changes. [27–29]

References

1. Aldon G, Panero M (2020) Can digital technology change the way mathematics skills are assessed? ZDM Math Ed 52:1333–1348. https://doi.org/10.1007/s11858-020-01172-8
2. Aldon AC, Schacht F, Swidan O (2021) Teaching mathematics in a context of lockdown: a study focused on teachers' praxeologies gilles. Ed Sci 11:382021. https://doi.org/10.3390/educsci11020038
3. Baber H (2021) Social interaction and effectiveness of the online learning–a moderating role of maintaining social distance during the pandemic COVID-19. Asian Ed and Develop Stud. https://doi.org/10.1108/AEDS-09-2020-0209
4. Bernard RM, Borokhovski E, Richard F, Schmid RF, Rana M, Tamim RM, Abrami PC (2014) A meta-analysis of blended learning and technology use in higher education: from the general to the applied. J Comput High Ed 26:87–122. https://doi.org/10.1007/s12528-013-9077-3191
5. Baber H (2020) Determinants of students' perceived learning outcome and satisfaction in online learning during the pandemic of COVID-19. J of Ed and E-Learn Res 7(3):285–292. https://doi.org/10.20448/journal.509.2020.73.285.292
6. Engelbrecht J, Borba MC, Llinares S, Kaiser G (2020) Will 2020 be remembered as the year in which education was changed? ZDM Math Ed 52(2):821–824. https://doi.org/10.1007/s11858-020-01185-3
7. Engelbrecht J, Llinares S, Borba MC (2020) Transformation of the mathematics classroom with the internet. ZDM Math Ed 52:825–841. https://doi.org/10.1007/s11858-020-01176-4
8. Sadovnichy VA Report of the Rector of Lomonosov Moscow State University, Academician of the Russian Academy of Sciences Viktor Antonovich Sadovnichy. Paper presented at the Congress of Teachers and teachers of Mathematics on 18–19 November 2021. Lomonosov Moscow State University. Available via DIALOG. https://event.msu.ru/congress/mct. Accessed 7
9. Bao W (2020) COVID-19 and online teaching in higher education: A case study of Peking University. Human Behav and Emerg Technol 2(2):113–115. https://doi.org/10.1002/hbe2.2022/02/08

10. Borba MC (2021) The future of mathematics education since COVID-19: Humans-with-media or humans-with-non-living-things. Ed Stud in Math 108:385–400. Available via DIALOG. https://doi.org/10.1007/s10649-021-10043-2
11. Heyd-Metzuyanim E, Aviv J, Sharon AJ, Baram-Tsabari A (2021) Mathematical media literacy in the COVID-19 pandemic and its relation to school mathematics education. Ed Stud in Math. Available via DIALOG. https://link.springer.com/article/https://doi.org/10.1007/s10649-021-10075-8. Accessed 08 February 2022
12. Silverman J, Hoyos V (eds) (2018) Distance Learning, e-learning and blended learning in mathematics education. ICME-13. Monograph. https://doi.org/10.1007/978-3-319-90790-1
13. Bonk CJ, Graham CR (2006) Handbook of blended learning: Global Perspectives, local designs. Pfeiffer Publishing, San Francisco, CA
14. Garrison R, Kanuka H (2004) Blended learning: Uncovering its transformative potential in higher education. Internet and High Ed 7(2):95–104
15. Graham CR, Jered B, Pulham E, Larsen R (2019) K–12 blended teaching readiness: model and instrument development. J of Res on Technol in Ed 51(3):239–258. https://doi.org/10.1080/15391523
16. Horn M, Staker H, Clayton CM (2014) Blended: Using disruptive innovation to improve schools. Education
17. Quinn D, Aarão J (2020) Blended learning in frst year engineering mathematics. ZDM Math Ed 52:927–941. https://doi.org/10.1007/s11858-020-01160-y
18. Alqahtani AY, Rajkhan AA (2020) E-learning critical success factors during the COVID-19 pandemic: A comprehensive analysis of Elearning managerial perspectives. Ed in Sci 10(9):216. https://doi.org/10.3390/educsci10090216
19. Octaberlina LR, Muslimin AI (2020) EFL students perspective towards online learning barriers and alternatives using Moodle. Google classroom during COVID-19 pandemic. Int J of High Ed 9(6):1–9. https://doi.org/10.5430/ijhe.v9n6p1
20. Buch GR, Warren CB (2017) The flipped classroom: implementing technology to aid in college mathematics student's success. Contemp Iss in Ed Res (Second Quarter) 10(2):109–115
21. Durak YH (2018) Flipped learning readiness in teaching programming in middle schools: Modelling its relation to various variables. J of Comp Assis Learn 34(6):939–959. https://doi.org/10.1111/jcal.12302
22. Dvoryatkina SN (2018) On the possibility and necessity of forming digital competencies by means of mathematics. Bull. of the Yelets State Univ named after IA Bunin 39:54–58
23. Graham CR (2006) Blended learning systems: Definition, current trends, and future directions. In: Bonk CJ, Graham CR (eds) Handbook of blended learning: Global perspectives, local designs. Pfeiffer Publishing, San Francisco, CA, pp 3–21
24. Graham CR (2013) Emerging practice and research in blended learning. In: Moore MG (ed) Handbook of distance education, 3rd edn. Routledge, New York, NY, pp 333–350
25. Heo H, Bonk CJ, Doo MY (2021) Enhancing learning engagement during COVID-19 pandemic: Self-efficacy in time management, technology use, and online learning environments. J of Comp Assis Learn 37:1640–1652. https://doi.org/10.1111/jcal.12603
26. Huang R, Ahmed TA, Wang H, Shi Y, Bonk CJ, Yang J, Burgos D (2021) Emergence of the Online-Merge-Offline (OMO) Learning Wave in the Post-COVID-19 Era: A pilot study. Sustainability 13(6):3512. https://doi.org/10.3390/su13063512
27. Iqbal J, Sidhu MS, Wang S (2017) A review on making things see: Augmented reality for futuristic virtual educator. Cogent Ed 4:1. https://doi.org/10.1080/2331186X.2017.1287392
28. Kelly D, Hoang TN, Reinoso M, Joukhadar Z, Clements T, Vetere F (2018) Augmented reality learning environment for physiotherapy education. Phys Therapy Rev 23(1):21–28. https://doi.org/10.1080/10833196.2018.1447256
29. Presmeg N (2014) Visualization and learning in mathematics education. In E Lerman (ed), Encyclopedia of mathematical education. https://doi.org/10.1007/978-94-007-4978-8

Teachers Training for STEM Education System

Training of Future Teachers for Blended Mathematics Education for School Students in a Digital Learning Environment

Natalia Podkhodova, Viktoriya Snegurova, and Vladimir Orlov

Abstract The modern difficult epidemiological situation has put forward new requirements for teaching mathematics to school students. Amid the pandemic they have to get their mathematics education remotely or through blended learning. Hence, students – future teachers of mathematics should be ready to organize such learning in this new digital learning environment. Therefore, the purpose of the study was to develop a model for training future teachers of mathematics for distance and blended mathematics education. The model is designed in the format of collaborative learning based on the interaction of five subjects of the educational space: teachers, masters and bachelors (future teachers of mathematics), university teachers and school students. The model was obtained from the analysis of previous studies, practical experience of distance learning, and theoretical analysis of the problem. The study tested the impact of collaborative learning (CL) on the level of readiness of future teachers for distance or blended learning in mathematics. The study used qualitative methods and statistical processing of the experimental results. The experiment showed that the differences in the level of training among the experimental and control groups for blended learning in mathematics turned out to be significant, which confirmed the efficiency of the model developed by the authors.

Keywords Digital learning environment · Blended learning of mathematics · Collaborative learning · Distance learning · Model of students' training

N. Podkhodova · V. Snegurova · V. Orlov (✉)
Russian State Pedagogical University, Named After A.I. Herzen, 48, Embankment of the Moika River St., St. Petersburg 191186, Russia
e-mail: vlvo@mail.ru

N. Podkhodova
e-mail: podhodova@gmail.com

V. Snegurova
e-mail: snegurova@bk.ru

A. L. Semenov et al. (eds.), *Structural and Technological Transformation of Education in the Post-Pandemic Period*, Lecture Notes in Networks and Systems 1282,
https://doi.org/10.1007/978-3-031-84039-5_14

1 Introduction

Distance learning was known long before the pandemic, starting from the eighteenth century. It was then that the "correspondence education" or "correspondence learning" appeared in the UK. It was used by Isaac Pitman, who taught shorthand to students. The bottom line was that students received all the necessary material, communicated with teachers and took exams using mail. The diversification of technical opportunities and means of communication has constantly expanded the functionality of distance learning. Besides, it existed in parallel and in a certain sense independent from traditional learning and required special training of teachers and was not widespread. Currently, distance learning is becoming an integral component of the information and learning environment of each educational institution. Each teacher can be involved in its implementation to various degrees. This situation was highlighted especially vividly due to the massive transition of educational institutions to distance learning amid the pandemic.

Russian schools and universities shifted to mass distance learning in March–April 2020. Not only schoolteachers and university teachers faced the problems of distance learning, but also students – future teachers who implemented this technology "from scratch" during their teaching internship at schools. At the same time, it was not only about the learning and classes themselves in a distance format, but also about various forms of extracurricular activities in mathematics. Moreover, as the epidemiological situation weakened, schools and universities again returned to offline learning, and then again switched to distance learning and back to offline. In fact, this was blended (hybrid) learning. In our study we understand blended learning as a change in the formats of offline and distance learning.

A digital learning environment is currently being formed in Russian schools and universities. Its creation, among other things, involves the use of various digital learning resources, but also the inclusion of various distance learning forms in the learning process. This allows putting forward the search for effective means of training the future teacher of mathematics for distance or blended learning of school students as a problem of study. This problem requires the purpose – to develop an effective model for students' training for distance learning in mathematics. This model should take into account modern trends in education and existing pedagogical experience, both offline and distance learning. As one of the foundations for the development of this model we chose the collaboration of subjects of the educational process: teachers, masters studying under the program "Mathematics Education", bachelors – future teachers of mathematics, teachers – internship supervisors and school students. Previously, the collaborative environment was used in Russian universities as a form of teaching students in specific academic disciplines [1, 2]. All the above confirms the relevance of the problem of the study.

The design, understanding and study of the efficiency of the model began in March–April 2020 at the beginning of the pandemic, when schools and universities in Russia massively switched to distance learning.

The rationale for the feasibility of building a model required the analysis of the experience and study of the problem of distance learning in our country and abroad.

1.1 Related Work

In the modern world the role of information and communication technologies (ICT), digital technologies are increasing, their use in different countries is becoming an integral part of the society. For example, in Japan, which is famous for technological progress, the intensity of integration of computer technology into education "significantly lags behind other highly developed countries. Therefore, Japanese online education is also lagging behind due to the low level of technology in Japanese higher education institutions and various cultural issues related to education and employment in the country" [3]. Besides, the use of the Internet by country residents does not particularly affect the attitude towards DL. According to the Portuguese National Statistical Institute, the level of Internet use by the adult population is about 76%. Among this population, people who receive secondary and higher education or have completed secondary and higher education show higher percentage of Internet use (98%). But as the study showed, only 5% of students in the Portugal group expressed great interest in distance learning courses [4]. The situation in the UAE is slightly better. About 100% of its residents use the Internet at home or on their mobile devices, but only 18% of students are strongly interested in taking DE courses. One of the reasons for this is the lack of a clear influence of the use of ICT in teaching mathematics on the efficiency of mastering this subject. Thus, 85% of Japanese teachers using these technologies noted that they did not find advantages in learning mathematics by students [5]. This also affected the low level of ICT in higher pedagogical education. There are also many universities that do not provide the DL opportunities because it is not part of their institutional culture.

Thus, even if students have gadgets and mobile phones, this does not mean that they have an interest in distance learning, i.e., students – future teachers should be motivated in the DL training. So, in a study by Keegan [6] distance learning is interpreted as an educational experience in which teachers and students are divided in time and space. Gunawardena, McIsaac, & Jonassen emphasize that DL is implemented away from an academic institution and may result in a degree or a diploma [7]. The key aspects of DL were presented in the paper [8]. These are elements such as the physical separation of a teacher and a student, learning in the context of an educational institution, use of technical media, communication between a teacher and a student, possibility of personal meetings, use of the industrial model of education. Various definitions of DL appeared in the twenty-first century, which are probably based on the experience of different countries, the point of view of various teachers and reflect the educational culture of these countries [4].

The most optimal conditions for the implementation of distance learning are being searched. To achieve this, many studies have been performed on various aspects of DL. Along with the DL opportunities and its widespread there are new studies highlighting not only the positive and negative aspects of DL courses, but also the challenges of distance learning [9]. The advantages of DL courses most often include independent learning, flexibility in time and space, time savings (lack of travel between home and school) and the fact that a distance learning course is often cheaper. Disadvantages include feelings of isolation, struggles with motivation, lack of personal interaction, difficulties in obtaining immediate feedback, need for continued and reliable access to technology, and sometimes some difficulties with accreditation [10–12].

If technological (or technical) problems can be solved, the training of future teachers in knowledge and skills also does not cause great difficulties, then the problems of applying these knowledge and skills in the real educational process in distance learning, as well as psychological and pedagogical problems are quite urgent in the training of students – future teachers. So the study [13] emphasizes that the problems of distance learning can be associated with the psychological state of students. In particular, this may include: a) problems caused by the lack of direct contact between a student and a teacher; b) problems related to the feeling of alienation and isolation from the student community; and c) concerns about the education and learning outcomes. The study [14] notes that eye contact is a very crucial factor in education but this factor can be limited in e-learning environment because lecturers may be unable to observe the emotions of students and cannot predict their satisfaction which puts burdens on them and make students respond differently toward e-learning.

There are the following types of online training: synchronous, asynchronous, mixed, massive open online courses (MOOC) and online courses with an open schedule. According to [15], live streaming video and/or audio is used for synchronous interaction. Although video conferencing allows participants to see each other, it is not considered a face-to-face interaction due to physical separation [8]. Asynchronous learning means that teachers and students do not have synchronous sessions and that students have access to course content online at any time they want or need. Communication between participants occurs mainly through email and online forums and is generally moderated by an instructor [15]. According to Garrison [16], "asynchronous coeducation may well be the defining technology of the post-industrial era of distance learning". Another type of DL is blended learning (BL). Thus, [17] define blended learning as a combination of personal offline classroom time with online learning experience. While it is unclear how much time is allocated for online in the blended model, "the real test of blended learning is the efficient integration of the two main components (offline and internet technology), so we do not simply add to the existing dominant approach or method". Different learning strategies and learning technologies can be used in BL format to help people who have different learning styles, needs and interests [18]. Another type of DL is MOOC (Massive Online Open Courses). This format was first introduced in 2006 and offers distributed open online courses that are available free of charge to a very large number of participants. By providing educational opportunities, MOOCs can meet the growing demand for training and education [19].

The training model we are developing is aimed at building the readiness of students – future teachers for distance learning, so it was necessary to analyze the literature on this issue. There are different tools and criteria for identifying the teacher readiness for professional activities. Many studies use a tool such as self-assessment of novice teachers for their professional readiness. So in the study conducted by [20]) teachers ranked their competence as "intermediate". The results of the study [21] show that teachers feel incompetent in the field of educational psychology, class management, assessment and some subject matters. The observation of practical school activities of students – future teachers by Gündüz and Odabaşı [22] made it possible to conclude that there was insufficient use of technology on the part of students in the real academic process.

Summing up the various definitions of this concept, we can say that most often collaborative learning means a certain educational approach to organizing education, in which groups of students work together to solve a problem, complete a task or create a product. This interaction between students and students and the teacher takes place through a dialogue (polylogue). Collaborative learning often occurs in a variety of contexts in order to address a particular problem [23]. As part of the interaction, the teacher acts not only as the organizer of joint education, but also at some point plays the role of the student in solving the problem.

2 Survey Methodology

The literature analysis made it possible to clarify the purpose and outline the study paths. So, the purpose of the study is to develop a model for training masters and bachelors – future teachers of mathematics to implement distance and blended learning in mathematics designed in the format of collaborative learning, and to test its efficiency. The study involved mathematics teachers of St. Petersburg, bachelor and master students of the Faculty of Mathematics of the Russian State Pedagogical University named after A.I. Herzen, faculty teachers delivering a series of methodological disciplines and supervising student' teaching internships.

The analysis of the literature made it possible to highlight the problems of students of foreign universities. It also showed that in different universities and different countries these problems have their own specifics. Therefore, at the beginning of the study, we conducted a survey covering teachers and students who underwent teaching internship in a distance format and students working at a school as the teachers of mathematics. It made it possible to specify the difficulties formulated in the introduction. As a result, we were able to highlight the main questions of teachers and senior bachelor students, master students in the implementation of distance learning:

1. How to choose forms, methods and means for online classes that are adequate to the new format?
2. What electronic educational resources in mathematics are available on the Internet? How can they be selected to achieve different learning goals? What are the criteria for their evaluation?

3. How to develop an online lesson scenario using different online platforms? (the lesson scenario is meant as the definition of the types of activities of students in the lesson taking into account the differences in conditions, the distribution of time for different types of activities, the forms of delivering the material, the methods used, etc.).
4. How to control students during an online lesson?
5. How to improve the efficiency and effectiveness of online lessons?
6. How to manage the cognitive activities of students in the process of distance learning?
7. How to choose the most efficient forms of interacting with students after school hours?
8. How to construct an individual trajectory for students to master the mathematical content?
9. How to manage the research and project activities of students in distance learning?

We see that most of the questions are related to the distance component of blended learning. Based on the analysis of the literature, the results of the survey and the experience described above, we developed a model for joint (collaborative) learning of students – future teachers of mathematics. The model describes a special form of interaction between five subjects of the educational space: mathematics teachers, masters, bachelors, consulting teachers (supervisors of teaching internships) and school students (Fig. 1).

Figures 1, 2 and 3 show that the model includes three stages of training for distance or blended learning. The first stage – information-acquainting – ensures the

Stage	Stage I. Information-acquainting
Function	Orientation
Content	Aimed at mastering knowledge and skills related to the specifics of distance and blended learning
Participants. Forms of collaboration	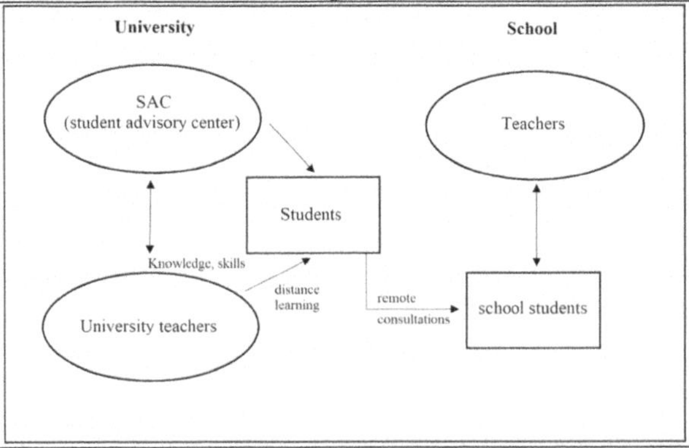

Fig. 1 Model of training students for distance or blended learning in a collaborative environment (Stage I)

Stage	Stage II. Training
Function	Search (test)
Content	Aimed at continuing training on the implementation of distance or blended learning and the primary implementation of developed materials for its organization (chats, consultations using educational platforms and information and educational resources)
Participants. Forms of collaboration	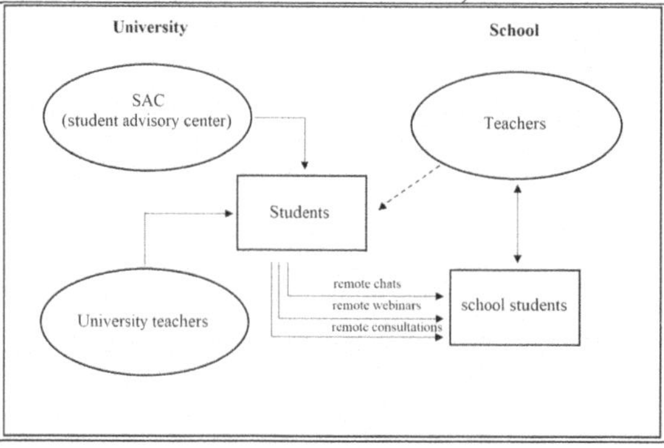

Fig. 2 Model of training students for distance or blended learning in a collaborative environment (Stage II)

orientation function. The second stage – training – ensures the testing function. The third stage – planning – ensures the design function.

Each of the five selected subjects within the model boundaries performs certain functions. The teacher, in accordance with the program and scheduling, formulates the topic of the lesson or a series of lessons to the student-trainee, in general terms helps to determine the subject content and the structure of the lesson corresponding to the didactic goal, based on which the student-trainee must design the lessons (from formulating the lesson topic and its goals, to the development of the scenario) and implement them. A student-trainee prepares a draft scenario highlighting the main stages of the lesson, selecting, and organizing theoretical and task material taking into account the specifics of a certain class and characteristics of students. Obviously, with this functioning of the model, the professional experience of all interacting subjects is enriched: student-trainee, student-consultant, teacher and supervisor both in terms of distance learning and in terms of traditional face-to-face interaction.

3 Results and Discussion

To check the efficiency of the developed model, it was necessary to identify the influence of the model on the level of preparation of students for distance or blended learning of school students in mathematics (formation of readiness). The teacher's

Stage	Stage III. Project-technological
Function	Planning
Content	Aimed at conducting lessons and after-hours lessons in a distance or blended format based on collaborative learning
Participants. Forms of collaboration	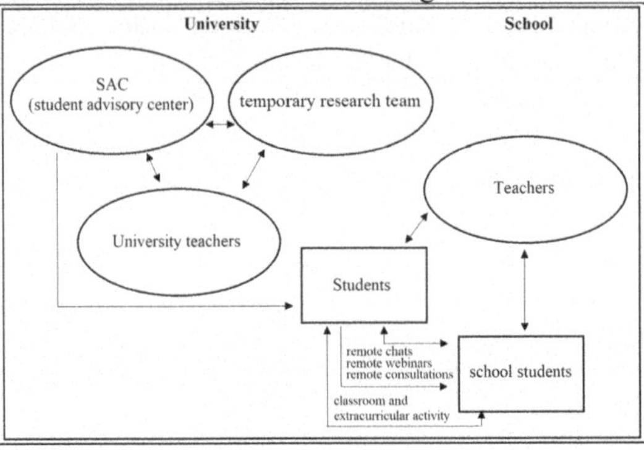

Fig. 3 Model of training students for distance or blended learning in a collaborative environment (Stage III)

readiness for professional activity includes three components: motivational, cognitive, and emotional-volitional [24]. During the experiment we studied the level of the cognitive component of readiness, because success in its formation increases the motivation of activity, and the emotional and volitional sphere of the future teacher is aimed at organizing training in any format.

To test the efficiency of the developed model as a means of preparing students for the distance (blended) mathematics education, we conducted an experiment during the teaching internship of bachelor and master students in the 2020–2021 academic year. There were two groups of students: control and experimental, amounting to, 42 and 56 people respectively, consisting of students of the third-, fourth-year bachelor and the first-year master students, who passed teaching internships at different times. All students were familiar with the classical teaching methods of mathematics and the basics of distance learning methods for mathematics, the use of information and communication technologies in teaching mathematics. Students of the control group passed teaching internships according to the traditional scheme, prepared distance learning in cooperation with the teacher and the supervisor, and students of the experimental group also had the opportunity to communicate with student-consultants who are part of the advisory center when preparing distance learning, as well as work as part of temporary research teams. The central issue in assessing the results was the cognitive component of the graduate's readiness for distance learning. The level of theoretical training (knowledge component) was checked using tests for knowledge of various aspects of distance learning. The tests that were offered to the participants in the control and experimental groups are given below. Test tasks:

Test 1

It is known that there are several groups of training methods during the training process, each of which is transformed in a certain way in the process of distance learning. Correlate the groups of training methods and the directions of their transformation. Verbal Visual Practical.

Test 2

Group the tools within the distance learning that correspond to each group of methods. Verbal Visual Practical.

Distance learning resource; Network teacher recommendations; Guidelines for network teacher; Video recordings of lectures; Animations; Computer models; Virtual excursions; Videos; Simulators; Workshops; Research models; Virtual labs.

Test 3

Correlate the selected advantages and groups of methods in distance learning. Verbal Vsual Practical.

Ability to repeat the same piece of content multiple times. Quick search for required information through hyperlinks. Study information is better absorbed through the comprehensive use of information presentation methods. Repeated practical action. Ability to quickly self-test and correct the action.

Test 4

Identify new means compared to traditional offline education, the use of which in the process of distance learning serves a factor in increasing its effectiveness. Remote resources; Computer animations; Forums and chats; Virtual labs; Computer simulators; Online computer testing systems; Video podcasts.

Test 5

Highlight the main specific conditions that affect the effectiveness of distance learning: Level of development of independent activities of students; Student motivation level; Teacher qualification level; ICT knowledge by a teacher; Readiness of a teacher and students for indirect interaction.

The level of practical training (skill component) was assessed through the methodological tasks of two types: checking the development of general methodological skills and skills related to the peculiarities of distance learning. The examples of such tasks are presented below. Example of the first group of tasks. Formulate the goals of studying the topic "Arithmetic and geometric progression" for classes with advanced study of mathematics. Make the scenario of the laboratory work on the topic "The sum of infinitely decreasing geometric progression".

Example of the second group of tasks. Analyze various digital educational resources on the topic "Creating polyhedron sections by plane". Select a visual series to learn how to create sections of a cube and pyramid using the trace method. Make up the text of the voice-over comment, which reveals the stages of plotting the section using the trace method. To check the effectiveness of the developed model of collaborative learning to form the readiness of graduates for distance or blended teaching of school students in mathematics, a numerical assessment of the results of the experiment was carried out. When performing these tasks in the control and experimental groups, students could score a total of 70 points (30 for tests and 40 for two methodological tasks). The tables of the results of experimental and control groups are given below (Tables 1, 2 and 3).

Table 1 Numerical results of the experiment

Experimental group		Control group	
Interval	Frequency	Interval	Frequency
[66;70]	10	[66; 70]	4
[61; 65]	22	[61; 65]	6
[56; 60]	4	[56; 60]	10
[51; 55]	15	[51; 55]	10
[46; 50]	5	[46; 50]	5
[41; 45]	0	[41; 45]	6
[36; 40]	0	[36; 40]	1
[31; 35]	0	[31; 35]	0
	$\overline{x} = 59.52$;		$\overline{y} = 54.64$

Table 2 Test results (knowledge)

Experimental group		Control group	
Interval	Frequency	Interval	Frequency
[28; 30]	10	[28; 30]	4
[25; 27]	22	[25; 27]	20
[22; 24]	23	[22; 24]	11
[19; 21]	20	[19; 21]	6
[16; 18]	0	[16; 18]	1
	56		42
$\overline{x} = 23.96$		$\overline{y} = 24.4$	

Table 3 Results on solving methodological problems (skills)

Experimental group		Control group	
Interval	Frequency	Interval	Frequency
[37; 40]	14	[37; 40]	4
[33; 36]	18	[33; 36]	6
[29; 32]	24	[29; 32]	14
[25; 28]	0	[25; 28]	11
[21; 24]	0	[21; 24]	7
	56		42
$\overline{x} = 33.79$		$\overline{y} = 29.45$	

The sample mean results in the experimental and control groups are as follows $\overline{x} = 59.52$; $\overline{y} = 54.64$. Let us compare the average of normally distributed general samples. To establish the significance of differences in average values in the experimental and control groups, we use the criterion $Z = \frac{\overline{x}-\overline{y}}{\sqrt{\frac{D_x}{n}+\frac{D_y}{m}}}$. The dispersions were $D_x = 41.75$; $D_y = 63.85$. The observed value of the test statistic was $Z_H = 3.25$, while the critical value for the significance level at $\alpha = 0.05$ was $Z = 1.64$. This allowed concluding that the differences in the average values of the experimental and control groups are significant, the level of readiness of the experimental group of students, in which collaborative learning for distance learning of school students in mathematics was conducted, was above the level of the control group.

We assessed the level of readiness of students for distance learning through knowledge and skills. Let us consider these components separately in the experimental and control groups (through tests and by solving methodological problems).

The dispersions were $D_x = 11.99$; $D_y = 8.13$. The observed value of the test statistics was $Z_H = 1.43$, while the critical value for the significance level at $\alpha = 0.05$ was $Z = 1.64$. The null hypothesis on the equality of averages is accepted. We can assume that the level of theoretical knowledge of the laws of distance learning in the experimental and control groups does not differ significantly.

The variances were $D_x = 10.54$; $D_y = 22.68$. The observed value of the test statistics was $Z_H = 5.11$, while the critical value for the significance level at $\alpha = 0.05$ was $Z = 1.64$. The null hypothesis of the equality of means is rejected. The level of skills in the distance learning in the experimental group is higher than the corresponding level in the control group.

The comparison of the results in Tables 2 and 3 suggests that with approximately the same level of theoretical knowledge of distance learning in the experimental and control groups, the skill component in the experimental group, in which collaborative learning was implemented, is much better, i.e., good knowledge of the theory is not a sufficient condition for qualitative implementation of distance or blended learning in practice.

4 Conclusion

The study made it possible to draw the following conclusions. Based on the analysis of domestic and foreign literature on the problems of distance learning, we have established that:

The ability of students to handle various gadgets, their use of various electronic communication media does not provide motivation for studying distance learning technologies and does not entail interest in distance learning as a component of blended learning in mathematics.

The distance learning of future teachers of mathematics need to be specifically trained at both theoretical and practical levels.

The training of future teachers for the use of modern information and communication technologies in teaching mathematics and various educational platforms should be strengthened both in the education of students in offline or blended format, and in the framework of distance learning.

The study confirmed that the designed model should provide for the interaction of students of the same year, students of various years, students and teachers in the process of offline studies at the university, as well as their interaction, and the collaboration of students with the teacher and students during teaching internships by implementing various forms of distance learning, i.e., be collaborative. This will remove the problem of a shortage of face-to-face communication during distance learning.

For the first time, we have developed and described a collaborative model for preparing students for distance and blended learning in mathematics during lessons and extracurricular activities (webinars and consultations, other forms of extracurricular classes in mathematics), which provides for various types of interaction between subjects of the mathematics learning process both on a permanent and temporary basis through the implementation of both horizontal and vertical training interaction between the main subjects of the mathematics learning process on a permanent and temporary basis through advisory centers and temporary research teams.

A local experiment showed that the implementation of this model increases the efficiency of preparing students for distance learning. It can be assumed that this model will improve the quality of methodological training of students and facilitate not only distance, but also offline learning in mathematics, enrich the methodological experience of practicing teachers, student consultants, motivate students and teachers of mathematics to use modern electronic educational resources in offline and distance learning. These hypotheses may determine the directions of further study in the professional training of future teachers of mathematics.

References

1. Medvedeva MS (2015) Formation of the readiness of future teachers to work in a blended learning environment. Dissertation Thesis. Nizhny Novgorod
2. Zaitseva ON (2013) Technology of distance learning in the process of organizing independent work of students of medical specialties of universities (by the example of studying mathematical disciplines). Dissertation Thesis. Orel
3. Nakamura M (2017) The state of distance education in Japan. Quarterly Rev of Distan Ed 18(3):75–87
4. Fidalgo P, Thormann J, Kulyk O (2020) Students' perceptions on distance education: A multinational study. Int J of Ed Technol in High Ed 17:18
5. Fullan M (2011) Choosing the wrong drivers for whole system reform. Centre for Strategic Education. Seminar Series 204
6. Keegan DJ (2002) The future of learning: From eLearning to mLearning. Central Institute for Distance University, Hagen
7. Gunawardena C, McIsaac M, Jonassen D (2008) Distance education. In: Jonassen D (ed) Handbook of research on educational communications and technology: Project of the Association for Educational Communications and Technology (AECT series). Lawrence Erlbaum Associates Inc., New York, pp 355–396
8. Keegan DJ (1980) On the nature of distance education. Central Institute for Distance University, Hagen
9. Mahlangu VP (2018) The good, the bad, and the ugly of distance learning in higher education. In: M Sinecen (ed), Trends in E-learning
10. De Paepe L, Zhu C, Depryck K (2018) Online Dutch L2 learning in adult education: Educators' and providers' viewpoints on needs, advantages and disadvantages. Open Learn 33(1):18–33

11. Venter K (2003) Coping with isolation: The role of culture in adult distance learners' use of surrogates. Open Learn 18(3):271–287
12. Lei SA, Gupta RK (2010) College distance education courses: Evaluating benefits and costs from institutional, faculty and students' perspectives. Ed 130(4):616–631
13. Pozdnyakova O, Pozdnyakov A (2017) Adult students' problems in the distance learning. Proc Eng 178:243–248
14. Usun S (2004) Factors affecting the application of information and communication technologies (ICT) in distance education (a case study of Turkey). Turkish Online J of Distan Ed–TOJDE 5(1):1–16
15. Watts L (2016) Synchronous and asynchronous communication in distance learning: A review of the literature. Quart Rev of Distan Ed 17(1):23–32
16. Garrison DR (2000) Theoretical challenges for distance education in the 21st century: A shift from structural to transactional issues. The Int Rev of Res in Open and Distan Learn 1(1):1–17
17. Garrison DR, Kanuka H (2004) Blended learning: Uncovering its transformative potential in higher education. The Internet and High Ed 7(2):95–105
18. Tseng H, Walsh EJ Jr (2016) Blended versus traditional course delivery: Comparing students' motivation, learning outcomes, and preferences. The Quart Rev of Distan Ed 17(1):43–52
19. Zawacki-Richter O, Naidu S (2016) Mapping research trends from 35 years of publications in distance education. Distan Ed 37(3):245–269
20. Üstüner M, DemIrtaş H, Cömert M (2009) The attitudes of prospective teachers towards the profession of teaching (the case of INONU university, Faculty of Education). Egitim ve Bilim 34:140–155
21. Gelbal S, Keleciolu H (2007) Teacher evaluation methods for measuring us and they met the proficiency perception problems. Hacettepe Univ, Faculty of Ed J 33:135–145
22. Gündüz S, Odabaşı F (2004) The importance of instructional technologies and material development course in the education of teacher candidates in the information age. Turkish Online J of Ed Technol 3(1):43–48
23. Smith BL, MacGregor JT (1992) What is collaborative learning? In: A Goodsell, M Maher, V Tinto, BL Smith, JT MacGregor (eds), Collaborative Learning: A Sourcebook for Higher Education. National center on postsecondary teaching, learning, and assessment publishing, Pennsylvania State University, USA
24. Kruchinina GA (1996) Didactic bases of formation of readiness of future teachers to use new information technologies of training. Dissertation Thesis. Moscow

Transdisciplinary Trend in the Transformation of the Content and Technologies of Teaching Mathematics and Computer Science to Future Teachers

Vladimir Testov, Evgeniy Perminov, and Oleg Golubev

Abstract Currently, in teaching mathematics and computer science to future teachers, there is an imbalance between the fundamentalization of their training and the rapid, insufficiently thought-out introduction of digital technologies into training. Imbalance was especially clearly manifested during the COVID-19 pandemic due to the lack of a clear and well-thought-out procedure for adapting trainees to the digital learning format. This article describes the leading role of the transdisciplinary approach in overcoming this imbalance, which is based on the study of such modern scientific fields as discrete mathematics, information theory, fuzzy set theory, big data and artificial intelligence. The importance of such transdisciplinary scientific fields for overcoming the emerging imbalance in education is also manifested in the fact that they allow the principle of a harmonious combination of fundamental and informational components of training to be implemented in the training of future teachers. Therefore, transdisciplinary scientific fields are crucial in raising the level of professional culture of future teachers, which is important for the implementation of multi-sided ties between all participants in the educational process.

Keywords Fundamentalization of education · Digital transformation · Discrete mathematics · Transdisciplinary trends · Imbalance in teachers' training

V. Testov · O. Golubev
Vologda State University, 2 Pobedy Ave, 160000 Vologda, Russia
e-mail: golubevob@vogu35.ru

E. Perminov (✉)
Ural Technical Institute of Communications and Informatics, 15, Repina Street, 620109 Ekaterinburg, Russia
e-mail: perminov_ea@mail.ru

A. L. Semenov et al. (eds.), *Structural and Technological Transformation of Education in the Post-Pandemic Period*, Lecture Notes in Networks and Systems 1282,
https://doi.org/10.1007/978-3-031-84039-5_15

1 Introduction

The ongoing digital transformation of education, accelerated by the COVID-19 pandemic, does not yet have a proper scientific base. The result of such a violation of the unity of education and upbringing was the destruction of multilateral ties between all participants in the educational process, which was especially adversely reflected in the implementation of the fundamental idea of unity of education, upbringing and development in pedagogy, which literally permeates the work of outstanding thinkers and teachers of the past.

The rapid digitalization of education has a particularly detrimental effect on the teaching of mathematics and computer science, which have always served as a model of accuracy and perfection. The achievement of this accuracy and perfection is significantly hindered by the avalanche-like spread of incorrect, non-standardized information technologies (IT) and developed sometimes by specialists far from mathematics and programming, guaranteeing a quick effect. In the organization of such a barrier in the form of IT standards that prevent pseudo-organizational attacks, it is extremely necessary to develop international recommendations and curricula that cause a high level of confidence in them in the professional environment and are the basis for standardization of the relevant educational activities of universities and universities [1, 2].

The negative impact of the hasty introduction of digital technologies into the teaching of mathematics and computer science occurs primarily due to the fact that the ideal, complex components of the phenomena of teaching mathematics and information (methodological, didactic, emotional) that exist only in the teacher's mind cannot be adequately digitized. Indeed, these components are stored in consciousness not in frozen "glazed" forms, and all their exact semantic meanings are best translated in live communication at lectures and other classes. Mathematics and computer science, as a rule, cannot be taught outside of "live" communication with a teacher who has a high level of pedagogical culture and skill.

Thus, the teacher is the key figure in overcoming the negative consequences of the digital transformation of education. In the training of future teachers, the importance of modern mathematics and information technology, which gave rise to the digital world and society, is great. Therefore, the problem of transformation of the content and technologies of teaching mathematics and computer science to future teachers in the new digital era is relevant.

In overcoming the imbalances of the digital transformation of the training of future teachers, the importance of the transdisciplinary trend in modern science is especially great, which brings scientific research to a deeper level of knowledge than the interdisciplinary one. Therefore, transdisciplinarity in science and education began to be called the locomotive of rapid economic development [3]. But it is no less important that transdisciplinarity "orients scientists to mutual enrichment with research experience gained in various disciplines" [4]. In the implementation of the transdisciplinary trend as a leading trend in modern education, such scientific fields as discrete mathematics, fuzzy set theory, information theory, catastrophe theory, big data, artificial intelligence, are especially important.

The main purpose of the article is to study the aspects of using the transdisciplinary trend in the transformation of the content and technologies of teaching mathematics and computer science to future teachers in order to find a certain balance between the fundamentalization of their training and its digitalization. In solving this problem, an important role is played by the implementation of the principle of a harmonious combination of fundamental and informational components of their training. This principle is especially important in overcoming the tendency to teach various recommendations for working with compelling instructional material in the context of commercialization and digitalization of education and the spread of algorithmic, less expensive learning technologies.

2 Research Methodology

The leading approach in the research methodology is aimed at the fundamentalization of the training of future teachers, implemented on the basis of identifying the role of transdisciplinary fields of mathematics and computer science in the modern digital world. These areas of mathematics have largely generated a transdisciplinary trend in modern science (called the transdisciplinary revolution in some works), which is of decisive importance for the implementation of advanced training. According to L. Florida, the transdisciplinary revolution marks the transition of humanity from history proper to hyperhistory [5].

However, this term has become widely used in the scientific literature in the last one and a half to two decades. Transdisciplinarity implies the emergence of various scientific systems that are above specific disciplines, above the disciplinary division of scientific knowledge, violation of the boundaries of various scientific disciplines.

As a result of the application of the new methodology, "such scientific fields as general systems theory, information theory, catastrophe theory, synergetics, artificial intelligence appear, which differ in principle by ignoring interdisciplinary boundaries" [6]. All these concepts play a leading role in rethinking the traditional functions of the teacher, as well as in defining his new "digital" functions.

In the transdisciplinary trend in modern science, the importance of modern mathematics is great, which underlies the harmonious combination of ideas and methods of various sciences in the conditions of their informatization [6, 7]. Therefore, the language of mathematics has the ability to combine different areas of theoretical knowledge, taking into account their specificity.

Therefore, the leading role in the study is played by the methods of fundamentalization of the training of future teachers based on transdisciplinary scientific fields of mathematics and computer science, essential for the improvement of the content and technologies of teaching mathematics and computer science.

The methods of this study are confirmed by the long-term experience of teaching students by the authors and their colleagues in various universities, as well as scientific research on teaching methods in interdisciplinary areas of mathematics, reflected in monographs [8, 9].

3 Results and Discussion

3.1 Mathematical Aspects of the Transformation of the Content and Technologies of Teaching

The basis for determining the goals, content and technology (methods, means and forms) of teaching an educational subject is always the external environment of the learning system. The process of digitalization of society in the digital era plays a leading role in the external environment of teaching mathematics and computer science to future teachers. In science and education, this process is closely intertwined with the transdisciplinary trend, the emergence of a universal methodology capable of solving complex multifactorial interdisciplinary problems of nature and society using the unique capabilities of modern computers. The foundation of this universal methodology is formed by transdisciplinary areas of modern science, especially mathematics and computer science. These are discrete mathematics, fuzzy set theory, information theory, catastrophe theory, big data, artificial intelligence, which affect or completely transform almost all professions, in particular, the teaching profession.

These areas are the scientific basis for using the unique capabilities of the computer in the transformation of professional activity. In the digital age, it is these areas that allow using the unique capabilities of a computer in almost any professional activity on the basis of unchangeable, universal mathematical knowledge, skills and abilities that do not depend on the conjuncture and time. Therefore, recently opened in some leading universities, the direction of training "Mathematics and Computer Science" combines the best that is in classical and modern mathematics and computer science.

As the analysis of the content of training in the mathematical foundations of computer science in leading Russian universities shows, abstract algebra, mathematical logic, combinatorics, theories of algorithms, automata and formal languages are of primary importance in such training. The ideas and methods of these scientific disciplines literally permeate the research of transdisciplinary fields of mathematics and computer science. Therefore, they are presented in one form or another in various textbooks on discrete mathematics, which is the basis of the language of information technologies.

One of the options for reforming the content of teaching mathematics and computer science in universities is to combine these sections into a single discipline "mathematical foundations of computer science". However, the ideas about the subject and functions of such a unified discipline differ significantly from different authors and are in the process In these conditions, it is necessary to develop generally accepted recommendations and curricula that are the basis for standardization of the content of this discipline in various universities. It is clear that the content of this discipline will be different for mathematical, technical, economic, humanitarian and other areas of training.

In the preparation of future teachers, such a combination of the listed sections into a single discipline allows them to be sufficiently represented in the content of teaching mathematics and computer science. In such training, it is important to implement the principle of a harmonious combination of the fundamental and informational components of their training. At the same time, in teaching the mathematical foundations of computer science to future teachers, it should be taken into account that the synthesis of ideas and methods of abstract algebra and mathematical logic has become the basis of the language of the theory of algorithms, automata and formal languages. In turn, the theory of formal languages is the basis of programming languages, database queries, grammars and recognizers [10], which are important in the in-depth study of computer science.

The mathematical foundations of computer science are of fundamental importance in the universalization of knowledge, skills, and skills in teaching mathematics and computer science based on the allocation of structural units of scientific knowledge, namely, mathematical structures and schemes (methods of cognition). Such structural units are algebraic, ordinal structures and logical, algorithmic and combinatorial schemes. The language of these structures and schemes literally permeates the language of the mathematical foundations of the digitalization of science and education.

On the basis of the listed structures and schemes in the process of teaching mathematics, mathematical cognitive structures and schemes are formed in a person's thinking, which are a reflection of these objectively existing mathematical structures and schemes. These mathematical cognitive structures and schemes underlie the correct use of computer technologies and computer mathematics systems in teaching mathematics and informatics, contribute to the "prevention" of errors of missed reasoning logic in the use of computer, hardware and software [8, 9]. It is equally important that they play a decisive role in the formation of the structure of intellectual operations, in overcoming the barrier between mathematical and other scientific knowledge [8].

As you know, the basis of almost all search engines are databases containing concepts and terms (keywords) and the addresses of sites where these keywords occur. It is important to take into account that the concepts of the language of mathematical foundations of computer science are the key terms of the thesaurus of the database of search engines in the field of mathematics and computer science. Therefore, these terms play a leading role in the innovative activities of future teachers and in their methodological preparation for the use of digital technologies in school.

Thus, the mathematical foundations of computer science are the basis for the implementation of the principle of a harmonious combination of the fundamental and information components of the profile training of future teachers to use the unique capabilities of information technologies of the digital era, especially such as artificial intelligence and big data. They are also the main factor in the fundamentalization of the profile training of future teachers in special courses in mathematics and computer science, designed to provide a transdisciplinary level of their training. Therefore, the mathematical foundations of computer science are crucial in overcoming the imbalance between the fundamentalization of the training of future teachers and the hasty process of digitalization of education.

3.2 Didactic Aspects of the Transformation of Content and Technologies of Teaching

Transdisciplinary concepts in the didactics of the digital era play a leading role in rethinking the functions of the teacher (constructive, communicative, organizational, stimulating, design, information). A new constructive function of the teacher has become especially important in the context of the widespread introduction of digital technologies into the educational process, since this function is associated with the analysis and design of the entire learning process, including the construction of its individual components: goals, content, methods, forms and means of teaching.

The implementation of this function is based on computer learning tools, which form the core of educational digital technologies. But in the context of the widespread use of digital technologies based on computer tools, the ideas that the interaction of a student with such means can allegedly be more effective than with a teacher and that computer tools and, in particular, training programs can largely replace a teacher in the learning process have become dominant. At the same time, it should be taken into account that the quality of the developed and used training programs leaves much to be desired, since most of these programs do not rely either on a recognized methodology or on important and proven theoretical developments. Therefore, in the didactics of the digital era, the principle of complementarity is fundamentally important, namely, information technologies should serve only as a supplement to the implementation of professional functions by the teacher. This principle is the didactic basis for ensuring a harmonious combination of fundamental and informational components of the training of future teachers.

Thus, the professionalism of the teacher, his personality are of decisive importance in the ability to conduct a deep analysis of the goals and objectives of teaching, designing content, choosing methods, forms of information technologies, especially computer-based learning tools. Therefore, in accordance with the principle of complementarity, it is the teacher who can select and use computer-based learning tools in an optimal way, taking into account the level of readiness of students.

It is important to take into account that the role of the teacher, the very content of his work in the conditions of digital transformation of education is changing significantly, especially in the conditions of the appearance (according to various sources) of at least 4,000 search engines used in different fields of activity. At the same time, no search engine (software package) will ever be able to collect and keep up to date all the necessary and complete information. Therefore, the teacher should be not only a source of knowledge, but also a navigator offering students the optimal learning trajectory for this course.

As already noted, the mathematical foundations of computer science are of great importance in the transformation of the content and technologies of teaching mathematics and computer science. Therefore, it is necessary to develop and implement a specialized course of training in these basics in the training of future teachers. Moreover, for the preparation of future teachers of mathematics and computer science, it is apparently necessary to introduce a special module consisting of a complex of such disciplines in order to train them as navigators in the "sea" of search engines.

There is already a solid didactic basis for selecting the content and technologies of teaching this course (module). Namely, as the analysis of textbooks and methodological literature shows, within the framework of traditional "pre-digital" didactics, didactic principles of teaching teachers abstract algebra, mathematical logic, combinatorics, the theory of algorithms, the theory of automata and formal languages have been fully developed [11, 12]. The content of such a special module should reflect the important main results of "quantification" studies obtained as a result of interdisciplinary interaction of mathematics, computer sciences and physics [13, 14]. In teaching these disciplines, the culturological principle, the principles of scientific, continuity, fundamental and professional-pedagogical orientation of the training of future teachers are especially important,

The key concepts of these disciplines form the main thesaurus of the database of numerous search engines in mathematics and computer science. Therefore, these concepts are important in the teacher's choice of the optimal trajectory of independent work of students for studying a particular course. As a result, students can independently supplement their knowledge with new theoretical concepts and developments reflected in scientific and educational literature.

4 Conclusion

As the study proves, transdisciplinary areas of mathematics and computer science are of fundamental importance in overcoming the negative consequences of the rapid, insufficiently thought-out introduction of digital technologies in the training of future teachers. They are the basis for the implementation of the principle of a harmonious combination of fundamental and informational components of teacher training, which is important in raising the level of their professional culture. Therefore, there is a need for a transdisciplinary expansion of the content of teacher training in the field of mathematics and computer science.

The theoretical and methodological basis for the implementation of this principle are such scientific fields as discrete mathematics, fuzzy set theory, information theory, catastrophe theory, big data, artificial intelligence, which have the greatest impact on the modernization of the training of future teachers in the digital era. In the digital age, it is these areas that make it possible to use the capabilities of a computer in almost any professional activity on the basis of unchangeable mathematical universal knowledge, skills and abilities that do not depend on the conjuncture and time.

The limitations of expanding the content of training future teachers, apparently, can be partially removed by using the unique capabilities of artificial intelligence and big data in teaching. Of great importance is also the new constructive function of the teacher, which underlies the analysis and design of the entire learning process, including the design of the content, methods, forms and means of teaching. As follows from the above, the main direction of expanding the capabilities of a teacher in the conditions of an avalanche of information and the use of digital technologies is his preparation for the use of specialized artificial intelligence systems that transform the necessary information into a visible data system necessary for a teacher to make didactic and methodological decisions. As a result, a teacher can be not only a source of knowledge, but also a navigator offering students the optimal learning trajectory for this course.

References

1. Computing Curricula (2020) Available via DIALOG. http://www.acm.orgbinaries/contentcurriculacc2020.pdf. Accessed 20 Feb 2022
2. SE2014 – Software Engineering (2014) Curriculum guidelines for undergraduate degree programs in software engineering. Available via DIALOG. http://www.acm.orgbinariesassets/education/se2014.pdf. Accessed 20 Feb 2022
3. Scholz RW, Steiner G (2015) Transdisciplinarity at the crossroads. Sustainability Sci 10(4):521–526
4. Pruzhinin BI (2015) Transdisciplinarity in the context of the dichotomy of applied and fundamental in science. In: Bazhanov V, Scholz RV (eds), Transdisciplinarity in philosophy and science: approaches, problems, prospects. Monograph. Navigator, Moscow, pp 252–262
5. Floridi L (2014) The Fourth Revolution. Oxford University Press, Oxford, How the Infosphere is changing human Reality
6. Testov VA, Perminov EA (2021) The role of mathematics in the transdisciplinarity of the content of modern education. Ed and Sci 23(3):11–34
7. Perminov EA, Testov VA (2020) Modeling methodology as the basis for the implementation of an interdisciplinary approach in the preparation of students of pedagogical directions. Ed and Sci 22(6):9–30
8. Perminov EA (2019) Methodical system of teaching discrete mathematics to students of pedagogical directions in the aspect of education integration. Monograph, 2nd ed. Russian State Vocational Pedagogical University, Ekaterinburg
9. Testov VA (1999) Strategy of teaching mathematics. Technological Business School, Moscow
10. Rosenberg G, Salomaa A (eds) (1997) Handbook of Official Languages, 1st ed. Springer, Berlin; New York, pp 1–3
11. Our MS (2020) Cellular automata based on the number system "Golden Mean" as the basis of artificial intelligence and artificial life. Int J of Artific Intellig and Mechatr 8(6):54–58
12. Thiruvengadam S, Tan JS, Miller K (2020) Artificial intelligence using hyper-algebraic networks. Neurocomput 399:414–448
13. Gyongyosi L, Imre S (2019) A survey on quantum computing technology. Comput Sci Review 31:51–71
14. Savchuk MM, Fesenko AV (2019) Quantum computing: survey and analysis. Cybern Syst Anal 55:10–21

Teaching Economic Disciplines Using Digital Technologies in Distance Learning

Yulia L. Yesina, Natalia M. Stepanenkova, and Maksim I. Shepelev

Abstract With the pandemic caused by the SARS-CoV-2 virus, the educational process has undergone a transformation, which has led to the need to search for new forms and methods of knowledge transfer. One of the key issues in this situation is the integration of humanitarian, economic, and digital knowledge into the technology of distance learning for international students. This article aims to study the prospects for the development of digital educational technologies and the main obstacles that cause difficulties in obtaining distance education by international students have been identified. The analysis of the educational process of students from the People's Republic of China at Russian University revealed several challenges in the transition to the use of digital technologies as a result of the social isolation that has arisen. The situation was aggravated by socio-cultural differences, language barriers, restrictions (both objective and legislative) in the choice of educational platforms and technologies, different levels of digital competence, as well as the difference in time zones. Generally, these factors reduced acceptable options for the rational organization of the educational process. This article shows that the construction of a strategy for the diversion of educational programs should be based on the personality-oriented teaching of students, providing an individual approach.

Keywords Distance learning · Digital technologies · International students · Economic disciplines

1 Introduction

On January 9, 2020, the World Health Organization (WHO) officially announced the discovery of a new coronavirus SARS-Cov2, which caused a surge in respiratory diseases COVID-19, which was declared a pandemic by WHO on March 11, 2020. The Chinese experience in combating the spread of the disease has shown the need

Y. L. Yesina · N. M. Stepanenkova (✉) · M. I. Shepelev
Bunin Yelets State University, 28, Kommunarov St, 399770 Yelets, Russia
e-mail: nmstepanenkova@yandex.ru

A. L. Semenov et al. (eds.), *Structural and Technological Transformation of Education in the Post-Pandemic Period*, Lecture Notes in Networks and Systems 1282,
https://doi.org/10.1007/978-3-031-84039-5_16

to introduce social distancing. The isolation measures taken immediately had a huge impact on the education system. According to UNESCO, since April 1, 2020, schools and higher education institutions (HEIs) have been closed in 185 countries, affecting 1,542,412,000 students, which is 89.4% of the total amount of enrolled students [1]. The full transition to distance learning was instantaneous and was accompanied by some difficulties for both students and teachers, administration, and support services [2].

The analysis of the implementation experience and the determination of the potential of distance learning gives an idea of new challenges in education, in solving them it is necessary to take into account psychological aspects, learning strategies, new formats, and platforms, modeling of the educational process, issues of confidentiality and ethics. The systematization of the conducted research shows that the quality of the educational process is determined not only by the availability of technologies but also by the effectiveness of applied socio-International student theories of learning, gamification, and emotional impact [3].

While close attention is currently being paid to the issues of using digital technologies in teaching, most studies focus on their role and practical results, such as student academic performance, omitting theoretical, methodological, or political aspects. Also, little attention is paid to the issue of social stratification, which has worsened and manifested itself in a sharper form as a result of the need for students to use digital educational technologies.

In this regard, the main trends in the international market of educational services were identified as research tasks, the study of factors determining the effectiveness of distance learning, including economic orientation and students' perception of the platforms and tools used. It is important to identify and systematize the features of the use of digital educational technologies in teaching economic disciplines to international students. To do this, it is necessary to analyze the digital technologies used for teaching economic disciplines in the higher education system, systematize practical experience, and determine the most optimal methods. The hypothesis of the study is based on the assumption that the systematization of experience in the implementation and expansion of distance learning tools, the introduction of new formats and platforms, and the adaptation of the educational process taking into account an individual approach and psychological aspects will smooth out problems in education and adjust the strategy of the university's behavior in the international market of educational services.

2 Survey Methodology

2.1 Trends in the Education Market

The development of digital technologies affects the global education market and the attractiveness and accessibility of its receipt. Thus, if in 2018 313.4 thousand international students were studying in the Russian Federation, then in 2021—324 thousand. The dynamics are influenced by factors such as the recognition of diplomas of education and the implementation of major joint projects by countries.

The general economic indicators and the level of the scientific school also have an important influence. The high level of Internet access and the presence of digital educational technologies makes it possible to provide educational services to Russian universities remotely. The low unemployment rate in the country makes it promising in terms of further employment of international students. Generally, in the Russian Federation, there are international students from neighboring countries – the countries of the Commonwealth of the Independent States, the Baltic States, and Georgia, which account for about 65%. However, there is a noticeable trend toward an increase in interest in learning from students from non-CIS countries, among whom a third of the students are Chinese students.

Many educational courses, including economic orientation, presented on this platform are free, but to obtain a specific profession according to certain educational programs or to obtain higher education in general, you need to pay. On the one hand, such opportunities open equal access to knowledge to people from anywhere in the world, anyone can listen to a course of lectures by leading scientists from around the world, and touch the world of advanced academic knowledge, but on the other hand, this is a rigid differentiation and differentiation of society into those who can afford to get an education in such universities in person and be part of the university environment, part of the scientific society and see the live work of the world's leading specialists, and those who do not have such an opportunity, but they are allowed to see part of this world through a computer or laptop screen.

This differentiation also applies to universities that can or wish to become participants in this educational platform. To create its online educational programs, the university needs to put a lot of effort into competing with the programs of leading universities already presented on this platform. Despite the diversity of universities participating in online platforms, applicants usually choose courses from those universities that are higher in international or regional rankings, that is, the problem of choosing a university for online learning is practically identical to full-time education.

Russia as a country of education is attractive to Chinese students of many preferences. Firstly, both Russia and China are major world powers with a common border. Secondly, the economic development of both countries was based on a common foundation of the political and economic system and an ideological base. Thirdly, the modern vector of transformations is socially oriented while using the tools of the market economy. Therefore, Chinese students do not experience discomfort and misunderstanding when communicating with Russian students and teachers. Taking into account the importance of developing continuing education programs in the context of the integration of Russian education into the world scientific community, an increase in international students is expected. Taking into consideration that mobile international students represent talented promising young people with the abilities and financial resources to enter and study at universities with high rating positions, tertiary education organizations significantly expand the set of programs for this category of students. At the same time, the strategy of diversification of educational programs is based on personality-oriented learning, providing an individual approach to the student to unlock the potential of each student.

2.2 Data and Methods

The system of research methods contributed to the achievement of the set goals. Their application made it possible to identify the dynamics of the studied object of research, to obtain effective indicators, and based on their objective assessment to draw appropriate conclusions and make a forecast of further development.

The methodology used is based on empirical methods aimed at actually occurring events. Thus, in the course of the study, through the study of various sources, the necessary information characterizing the object and the subject of the study was collected and systematized. Using various comparisons, the locality, reliability, and relevance of the collected material were established. When conducting the survey, a statistical aggregate was formed, and factorial signs and evaluation criteria were determined.

Quantitative methods, primarily statistical ones, made it possible to obtain real estimates and present them with a rating scale of a Liker, to establish existing patterns using rank correlation coefficients (for example, Spearman). The graphical method reflected the observed dynamics and identified development trends. The bibliometric method made it possible to compare the results obtained with the opinions reflected in other, both national and foreign scientific works.

To achieve this research goal, using the example of the experience of teaching economic disciplines in a specific organization of higher education, the attitude of students to the use of digital technologies and the use of distance learning tools during online classes in economic disciplines was studied. At the same time, the questionnaire method was used during the session falling during the pandemic period. The statistical aggregate was made up of the educational institution of the Federal State Educational Institution «Bunin Yelets State University» (students of economic profiles). The respondents were 91.67% of the total amount of foreign students studying in the economic profile using digital technologies. Of the total amount of completed questionnaires, 3.2% were rejected for various reasons (incomplete or incorrectly filled out).

3 Results and Discussion

Various approaches are used to evaluate the effectiveness of using distance learning tools. In many works, methods of grouping, classification, and structural analysis were used [4].

The issues of rational and effective organization of the process of distance learning are based on the use of psychological methods for studying students' self-regulation skills and metacognitive abilities when conducting classes online [5]. The method of in-depth interviews is often used in studies of this kind. The use of a questionnaire on the Likert scale has been recommended as an effective method.

During the study, a separate educational institution of Yelets State University was selected as a statistical aggregate. Currently, 232 international students receive education in bachelor's and master's degree courses here. At the same time, 55 students are located on the territory of the Russian Federation, and 117 international students receive education remotely. The conducted research on students' attitude to the process of obtaining knowledge in economic disciplines under the conditions of restrictions caused by the pandemic and the use of distance learning tools showed a rather tense attitude of respondents to the exclusively online form. The conducted surveys show that students whose material position allowed them to have available free rooms, communication devices with the necessary technical parameters, and also their parents' education, their free time, and digital skills were in a more advantageous position.

Systematization of the survey results showed that 58.3% of respondents hope to receive permission to travel to the country of study to continue their education offline. A high proportion of students (24.6% of respondents) are interested in continuing the distance learning method and are not going to leave their country of residence. A separate group (17.1% of respondents) includes those who have not decided on their attitude to the situation, as well as those who are going to postpone getting an education (8.6%) and intend to switch to work (4.7%), search for vacancies and internships (2.2%) or study courses on the online economy (1.6%).

At the same time, the survey found that, despite the apparent convenience of obtaining higher education at home, without spending money on paperwork, moving, and living in another country, the majority of respondents would prefer to study at a university campus. Respondents noted that they were counting on the temporary nature of learning using digital platforms and hoped to gain experience in full-time education.

To assess the closeness of the connection and the qualitative characteristics of the studied statistical population, Spearman's rank correlation coefficient was calculated.

Since the Spearman coefficient takes into account the difference only of ranks, and not of the values of the aggregate themselves, it does not accurately reflect the presence and degree of functional connection. Therefore, the t-criterion (critical point) was used to assess the statistical significance of the obtained coefficients. His calculation showed that for all factor values, the coefficient value is less than the level of the critical point. This indicates the statistical significance of the rank correlation coefficient. These conclusions were confirmed by the Chaddock scale.

The use of digital educational technologies in various volumes shows different results [6–8]. The level of development of the necessary infrastructure, the availability of devices that support educational platforms and the workplace of the student for distance learning differs significantly.

Thus, universities in the Asia–Pacific region are 99% equipped with the necessary equipment and infrastructure, North and South America and Europe—97%, while universities in Africa – are only 66%. In other words, 34% of African universities lack communication infrastructure.

As a result, the following main conclusions can be formulated, which are naturally traced in all studies concerning online learning and the use of digital distance technologies.

The advantages of online learning include the following main aspects:

Firstly, digital distance technologies make education accessible to a majority of students. Moreover, this accessibility means both physical and temporary one, that is, to receive education, it is necessary to have a minimum set of technical means of communication, which makes it possible to get acquainted with the material at a convenient time for the student. But this is only possible if lectures and other types of classes are available on the record in the form of a file, which is not always possible.

Secondly, the student can create conditions for himself that contribute to better assimilation of the material, and concentration on the learning process.

Thirdly, digital contacts with classmates contribute to the rapid exchange of information and help to keep up with the learning process and not miss anything.

Fourth, from the point of view of the teacher, it is very convenient to use online for practical work in small groups of up to 7–10 people. The advantage is that it is possible to share material, and visual tools in real-time, give tasks to be completed and control the process of the student's work on the task received, as well as discuss the progress and results of the work both individually and in the group as a whole.

Fifthly, students of IT specialties see an unambiguous plus in online training, since, in the traditional format of training, IT specialties involve the active use of digital information and communication technologies both in the lecture process and in practical classes. And from this point of view, such training in this field is perhaps even more preferable to the traditional one.

For a teacher, the disadvantages of online technologies are as follows.

Firstly, it is almost always difficult to deliver a lecture online, because the lecturer very often does not see the audience, does not immediately see the feedback of students, and cannot visually determine the degree of perception of information by the audience.

Secondly, taking into account the first negative factor, for many teachers, the lecture turns into a monotonous presentation of the material. Psychologically, this demoralizes the teacher and reduces the incentive to further work. 58% of teachers have difficulties with increased time spent preparing for classes in an online format.

Thirdly, online teaching often demotivates and destroys the personality of the presenter, since there is a great temptation to relax and not take a very responsible approach to lectures since there is no feedback and it is possible to replace most of the lectures with interactive materials or materials from the Internet, that are freely available.

It is impossible not to note the paradox of today, which consists in the fact that despite the democratic and practically accessible way of obtaining education or individual competencies online for many people anywhere in the world, it nevertheless turns into a tool that contributes to the expansion of inequality both among individual students and among universities in general.

As for students, firstly, the technical capabilities to participate in online learning are completely different for everyone. Someone has a separate room for this with a stationary computer or laptop, with high resolution, sound, and other technical parameters, and someone can listen to lectures only with the help of a phone in conditions that do not contribute to the perception of the information received [5].

Secondly, universities have always been and remain a cultural center, whether it is a metropolis or a provincial city. And universities can only be such a center offline, when students visit them and interact with teachers, with each other, and with invited specialists, this is part of the student's cultural society. Thus, the social stratum of society, called the intelligentsia, is formed – all this is a long social process of personality formation and its further adaptation both to the professional sphere of future activity and to society, including the formation of the skills to communicate, to seek compromises, to perceive the surrounding reality with tolerance and to organize their work in large and small groups.

Thirdly, in the process of full-time study, interpersonal connections are formed among students in the educational environment, which then determine the trajectory of the student's further life after graduation, since later classmates turn into colleagues, business partners, and sometimes into bosses or subordinates.

Fourth, if we are talking about full education through online technologies, through the organization of training on appropriate platforms, then a significant disadvantage is the fact that for an organization selling online courses, in the end, it does not matter whether a student can realize himself after training or not, the organization is interested in profit, that is, some people who will want to study according to their programs. [9, 10]

In addition, our analysis of the experience of teaching Chinese students in the Master's degree program showed a few other issues that are fundamental in the online learning system, and overcoming them even online causes certain difficulties. In particular, China's economic development model is peculiar and based on market relations with the leading role of the public sector and under the serious ideological control of the Communist Party. And this is a relatively new phenomenon in world economic history. In this regard, the teaching of classical economic disciplines, such as micro and macroeconomics, public finance, and the state budget system, faces a completely different economic and ideological perception, and sometimes resistance. The fact is that the political and economic liberalism of Western economic theory, which is generally recognized as an academic university discipline, can have serious problems in competition with the new socio-economic system.

4 Conclusion

The assessment of trends in the education market showed an ambiguous attitude of various scientists studying this issue, as well as students and teachers to the transition to a digital learning format. On the one hand, the use of digital technologies for admission and education has expanded the possibilities of studying in universities in other states. On the other hand, the abrupt transition to distance learning revealed a few difficulties associated with both the level of digitalization and the skills of applying new technologies and the reluctance to depart from traditional forms. However, the pandemic caused by the SARS-CoV-2 virus has dramatically accelerated the development of distance education methods. The most technologically advanced and prepared for this transition were the USA, Great Britain, Germany, China, and Japan. The global online education market is estimated at $165 billion with an annual growth of 5–15% per year. At the same time, the high level of access to the Internet, and the availability of digital technologies in the Russian Federation make the country competitive in the market for educational services. International students from neighboring countries study mainly in the Russian Federation, which accounts for about 65%.

A system of methods contributed to the achievement of the research goal. By the example of teaching economic disciplines in a specific organization of higher education, the attitude of students to the use of digital technologies and the use of distance learning tools was studied using the questionnaire method. The assessment of the closeness of the connection and the qualitative characteristics of the studied statistical aggregate was confirmed by the value of Spearman's rank correlation coefficient and the Chaddock scale.

All the pros and cons of online education and distance learning with the use of digital technologies identified in the Results and Discussion section are typical for all universities and all areas of training, except for IT technologies, including the teaching of economic disciplines. The example of the Bunin Yelets State University for the preparation of master's in finance and credit among students from the Chinese People's Republic confirms all the previously obtained conclusions about the pros and cons of online learning. Even while studying abroad, Chinese student representatives try to stipulate important, in their opinion, ideological points in the teaching of disciplines. This, on the one hand, limits the ability to give examples of economic relations and demonstrate the operation of economic laws in a particular area, based on the example of leading countries and companies in the world economy. On the other hand, the study of economic disciplines by Chinese students in Russian universities is much easier in terms of adapting the material of disciplines, since the Russian economic system, albeit not fully, is still similar to the Chinese one, and according to certain aspects of economic interaction of economic entities, it is quite possible to draw analogies in the Russian and Chinese economy.

References

1. Marinoni G, Van't Land H, Jensen T (2020) The impact of Covid-19 on higher education around the world. IAU Global Survey Report. Available via DIALOG. https://www.iau-aiu.net/IMG/pdf/iau_covid19_and_he_survey_report_final_may_2020.pdf. Accessed 24 Apr 2022
2. Bao W (2020) COVID-19 and online teaching in higher education: a case study of peking university. Hum Behav and Emerg Technol 2(2):113–115. https://doi.org/10.1002/hbe2.191
3. Abbas A, Fatima A, Arrona-Palacios A, Haruna H, Hosseini S (2021) Research ethics dilemma in higher education: Impact of internet access, ethical controls, and teaching factors on student plagiarism. Ed and Inform Technol 26. https://doi.org/10.1007/s10639-021-10595-z
4. Holmes W, Nguyen Q, Zhang J, Mavrikis M, Rienties B (2019) Learning analytics for learning design in online distance learning. Distan Ed 40:1–21. https://doi.org/10.1080/01587919.2019.1637716
5. Mustafa N (2020) Impact of the 2019–20 coronavirus pandemic on education. Int J of Health Prefer Res 5:31–44. https://doi.org/10.1016/S14733099(09)70176-8
6. Badiozaman IFA (2021) Exploring online readiness in the context of the COVID-19 pandemic. Teach in High Ed 28(5):1–19. https://doi.org/10.1080/13562517.2021.1943654
7. Aristovnik A, Keržič D, Ravšelj D, Tomaževič N, Umek L (2020) Impacts of the COVID-19 pandemic on the life of higher education students: a global perspective. Sustainability 12(20):8438. https://doi.org/10.3390/su12208438
8. Coffey J, Cook J, Farrugia D, Threadgold S, Burke PJ (2020) Intersecting marginalities: international students' struggles for "survival" in COVID-19. Gender Work Organiz 28(4):1337–1351. https://doi.org/10.1111/gwao.12610
9. Afacan Adanir G, Muhametjanova G (2021) University students' acceptance of mobile learning: a comparative study in Turkey and Kyrgyzstan. Ed and Inform Technol 5:6163–6181. https://doi.org/10.1007/s10639-021-10620-1
10. Kampylis P, Punie Y, Devine J (2015) Promoting effective digital-age learning—a European framework for digitally—competent educational organizations. https://doi.org/10.2791/54070

Digital Transformation of the Procedure for Assessing Competencies in a Pandemic

Victor N. Gusyatnikov, Tatyana N. Sokolova, Aleksey I. Bezrukov, and Inna V. Kayukova

Abstract The pandemic shift to distance learning has revealed a lack of reliable tools for online assessment of learning outcomes. Existing distance learning platforms do not provide effective tools for mass and objective assessment of competences. This paper proposes the use of an intelligent assessment system that combines the massiveness of computer-based testing and the personalised approach of an oral examination. The aim of the research is to develop an adaptive methodology for assessing the level of students' competence formation in one testing session. The methodology is based on the modernised Rasch model and uses a Bayesian algorithm to classify students according to the level of competence formation. After each task, the probability of a student belonging to certain partners (sets of values of formed competences) is calculated. In order to speed up the estimation process, an adaptive job selection is used, oriented to maximise the information function for the partner whose probability of affiliation to the previous step was maximal. To improve the stability of the Bayesian algorithm, a regularisation method based on the analysis of the change in the entropy of the distribution of the probability of belonging to partners is used. In case of atypical entropy change, it is suggested to re-run the task with the same difficulty parameters. This is analogous to asking a clarifying question during a face-to-face exam, which increases the robustness of the Bayesian algorithm. The simulation experiment has shown that the developed model allows us to reliably measure the level of competences formed during one testing session by performing several tasks.

Keywords Computer-aided assesment · Rasch model · Competence approach · Adaptive testing · Distance learning

V. N. Gusyatnikov (✉) · A. I. Bezrukov · I. V. Kayukova
Yuri Gagarin State Technical University of Saratov, 77, Politechnicheskaya Street, 410054 Saratov, Russia
e-mail: victorgsar@rambler.ru

T. N. Sokolova
Saratov State Law Academy, 104, building 3, st. them. Chernyshevsky N.G., 410056 Saratov, Russia

A. L. Semenov et al. (eds.), *Structural and Technological Transformation of Education in the Post-Pandemic Period*, Lecture Notes in Networks and Systems 1282,
https://doi.org/10.1007/978-3-031-84039-5_17

1 Introduction

The COVID-19 pandemic has posed an unprecedented challenge to education systems around the world, causing the most widespread disruption in history. More than 1.6 billion students in more than 190 countries on all continents were affected by this global problem. The closure of schools and other educational institutions affected a huge number of students–94 per cent of the global total. The pandemic has had a particularly strong impact on low- and lower-middle-income countries, where school closures have affected 99 per cent of students. This situation has caused many problems in education. School closures have led to the cessation of regular classes and disruption of the educational process. Many students lost access to education and were unable to continue their studies. Not all students had equal access to distance learning. In low-income countries, lack of access to the Internet and computers made distance learning unaffordable for many students. School closures and isolation from friends and teachers had a negative impact on the psychological health and well-being of many students. The COVID-19 pandemic has highlighted the vulnerability of education systems and the need to develop more resilient and adaptive approaches to learning. It is important to address the needs of all learners and ensure equal access to quality education in all settings. The massive transition of the education system during the pandemic to a distance format has revealed serious problems in assessing learning outcomes in many national educational systems, including the Russian one. As it turned out, with a wide variety of online learning platforms in the education system, there is no reliable toolkit for remote, mass and objective assessment of the level of competencies formed in the learning process. That is, the traditional system of monitoring and assessing the quality of education was not able to perform all its functions in a pandemic, which led to the postponement and even cancellation of final attestation in many national educational systems [1]. The modern view of the quality of education moves away from the traditional understanding as a simple set of knowledge and moves to a more comprehensive approach. The quality of education is seen as the degree of compliance of learning outcomes not only with established standards, but also with the real needs of the learner, as well as with the planned learning outcomes. This means that the quality of education is not determined only by the amount of knowledge acquired in specific disciplines, but also by the level of formation of a certain set of competences of the learner. Competences are a set of knowledge, abilities and skills that allow the learner to solve practical tasks and successfully function in the environment. Therefore, the means of control of students should be able to assess not only individual knowledge, but also the level of formation of several competences at the same time. This will allow to get a more complete and realistic picture of students' progress. The purpose of this paper is to develop an adaptive methodology for assessing the level of several competences of students based on the results of a single testing session. This methodology should allow real-time adjustment of individual characteristics of the learner, which will make the evaluation process more flexible and effective. This approach will allow to take into account the individual pace and learning styles of each student, as well as help to identify gaps in knowledge and skills in order to then focus on their elimination.

2 Survey Methodology

The problem of assessing the level of competency formation, which arose with the transition of education to a competency-based model, is a key one and has not been fully resolved for modern Russian education [2]. The task of assessing the achievements of students in a certain discipline, which arises at all stages of education, is even more complicated, since each discipline, as a rule, forms from two to four or more competencies. Therefore, the task is set in such a way that during the intermediate certification during one assessment procedure it is necessary to determine the level of formation of all these competencies. It is impossible to solve such a multidimensional problem of competency assessment using simple linear algorithms for analyzing test.

On the other hand, the use of complex methods for analyzing test results with elements of artificial intelligence [3] gives rise to another problem due to the insufficient level of trust in such assessment systems, both on the part of students and teachers. The problem of trust in assessment systems escalated precisely during the pandemic, when the widespread use of distance learning technologies became a necessity.

The testing technique proposed in this paper is a natural generalization of the Rasch model. It is assumed that the correct performance of the task requires a certain level of development of several competencies. The problem is to assess the impact of each competence on the probability of a correct answer. As is known, in the classical Rasch model, this probability is described by a logistic curve and depends on the difference between the difficulty of the task δ and the level of preparedness of the subject θ.

The main hypothesis of this work is that in order to estimate the probability of completing a task, it is necessary to choose the maximum probability.

To illustrate the proposed approach, consider the following model example. Let the subject be given the task: to cross a wide river. To complete the task, he can swim across the river, showing the art of a swimmer, build a raft, showing skill in its construction, or negotiate with a boatman, showing the skill of a negotiator. Thus three competencies are required to complete a task, albeit to varying degrees. However, if the test subject decides to build a raft, then it does not matter how well he can swim or negotiate.

The following assumptions are used in the proposed model: firstly, when performing a task, the test subject, trying to show the best result, uses exactly the competence that allows him to perform it with the greatest probability. Secondly, each competence is manifested independently of the others and can be assessed using the Rasch model.

Using the above assumptions, we can apply various testing algorithms:

- classical testing, in which the subject is asked to complete a certain set of tasks, and the result is evaluated by the maximum likelihood method [4, 5];
- a deterministic Bayesian algorithm, in which after each execution of a task taken from a fixed sequence, the probabilities of the subject belonging to each pattern are recalculated [6];
- adaptive Bayesian algorithm, in which each next task is selected taking into account the result of the previous tasks [7].

In order to test different testing algorithms and compare their characteristics, a simulation model of the computer-based testing process was developed. The model includes 'students' with different levels of three competences (θ1, θ2, θ3) and 'test tasks' with different levels of complexity for each competence (δ1, δ2, δ3). The model establishes a correspondence between the assessment scale and the level of competence formation measured in logits: excellent–3, good–1, satisfactory–minus 1, unsatisfactory–minus 3. To take into account all possible combinations of student's preparedness and task complexity, sets (θ) and (δ) corresponding to certain patterns are created. In the simplest case, there are 64 types of 'students' and 64 types of 'tasks'.

The testing process is simulated as follows. For each pair "student"—"task" for each of the three competencies, the probability of successful completion $P_{n..}$ is determined and the probability of task completion is calculated as max(P_1, P_2, P_3). The quality criterion of the method is the degree of coincidence of the true (specified in the model) levels θ_1, θ_2, θ_3 and the estimates of these characteristics obtained by the method under study.

We can assign an information function to each pair of "student"—"task" [8]. The value of the information function will be maximum for the task, the result of which carries the maximum information about the level of competence formation of this "student". In the simplest case, the information function is calculated as the product of the probability P_{ij} of successful completion of the *i-th* task by the *j-th* student by the probability of failure (1- P_{ij}) of this task. The values of the information function of the task are calculated as the sum of the information functions for all types of students:

$$I_i = \sum_{j=1}^{64} P_{ij}\left(1 - P_{ij}\right) \tag{1}$$

In the classical testing algorithm, the result is determined by the maximum likelihood method [9, 10]. For the resulting list of task execution results, a pattern is selected for which the likelihood ratio is maximum.

In a deterministic Bayesian algorithm, the "student" is also presented with a certain set of tasks. The task of evaluating the competencies of a "student" is reduced to the task of determining the pattern to which the student belongs, i.e. to the classification problem [11, 12].

Before the first task, it is assumed that the probability of belonging to each pattern is equal to 1/64. This means that all partners are assumed to be equally likely before testing. After each task, the probability of belonging to each pattern is recalculated using the Bayes formula. This formula takes into account the outcome of the task and redistributes the probability of belonging to the partners. As a result, the student being tested belongs to the pattern that has the highest probability of belonging to it. In contrast to the deterministic approach, in the adaptive Bayesian algorithm each next task is selected from the list of tasks not yet used so that its fulfilment or non-fulfilment gives the maximum information about the pupil's belonging to one of the patterns. This is the task for which the value of the information function is maximised. This allows the testing process to be optimised by selecting tasks that are maximally informative for assessing the student's competence [13, 14]. The algorithm ends when the probabilities of belonging to partners stop changing significantly after the next task is completed. This means that the evaluation system has received enough information to determine the student's competence level. Thus, the algorithm mimics the behaviour of an experienced examiner who asks several questions to assess the student's level of competence. The adaptive approach allows a more efficient and accurate assessment of students' competences in the process of computer-based testing.

3 Results and Discussion

In the simulation model, a student with a given level of development of each of the three competencies C_1, C_2, C_3 was tested. The probability of a given student's correct answer to the *k-th* question with difficulty $(\delta_1, \delta_2, \delta_3)$ was determined as the maximum of the probabilities:

$$P_k = maxP(C_n, \delta_n) \tag{2}$$

where n = 1, 2, 3—serial number of competences.

Further, on the basis of this probability P_k, the result of the assignment by the student is modeled. If a uniformly distributed random number $r \epsilon (0,1)$:

$$r \leq P_k \ task\ completed; \ r > P_k \ task\ not\ completed.$$

The probabilities of a given student belonging to each pattern were recalculated after receiving an answer to the next question. The type of the *k-th* question (m) was chosen based on the maximum value of information function. A good indicator of the process of clarifying the subject's belonging to patterns is the entropy of the distribution of their probabilities, calculated using the Shannon formula.

Simulation experiments show that in the case of an adaptive question selection algorithm, the probability that the subject belongs to the pattern set for him grows rapidly and reaches a maximum value of more than 0.7 by the twentieth task. For a deterministic question selection algorithm, this probability grows much more slowly and does not exceed 0.4 even after the fortieth task. The value of entropy in the first case decreases by the twentieth task by three times (from 5.7 to 1.7), and in the second case by the fortieth question less than twice (from 5.9 to 3.2).

The application of the Bayesian method for assessing the level of competence formation faces the problem of low stability of the algorithm to random answers of the test taker. In this system, each task is selected taking into account the maximum information function for the student's current level of competence. This means that the probability of a correct answer to each question will be close to 50%. As a result, when the same student is tested again, different answers to the same types of questions are observed, which affects the results of measuring the probabilities of pattern membership. The measurement results are particularly affected by the answers to the first 10–15 questions in the test. Simulation experiments show that depending on the answers to the first 15 questions, the Bayesian probability can change under different scenarios, while subsequent questions have little effect on the final measurement result. The reason for the instability lies in the peculiarity of the Bayesian algorithm, which becomes overly sensitive to random factors affecting the outcome of the task. It is important to note that similar problems arise in traditional exam taking. For example, a student may accidentally answer a complex question correctly or get a simple question wrong. This can distort the assessment of the student's level of knowledge.

The analysis of the behaviour of the entropy of the probability distribution makes it possible to improve the algorithm's robustness to random influences. Figure 1 shows that sharp changes in the probability of a student belonging to a certain pattern are accompanied by 'atypical' entropy changes. For example, a sharp decrease in probability after the thirteenth and sixteenth question is accompanied by an increase in entropy, although normally the entropy of the probability distribution should decrease after each completed task. The same sharp decrease in probability after the next answer may be accompanied by a sharp decrease in entropy. Thus, a sharp decrease or increase in the entropy of the probability distribution by partners after a task is completed may indicate that something is wrong with the obtained answer. To regularise the obtained solutions, the following procedure is proposed: if the entropy of the probability distribution over the patterns decreases (by more than 20% of the current value) or increases sharply after the next task, the result of this task is not taken into account when calculating the Bayesian probability. Instead, the student is offered the next task with the same difficulty parameters as the one just completed.

This allows to refine the test results and smooth out the influence of random answers on the final grade. The Probability2 and Entropy2 curves in Fig. 1 show the results of applying the proposed regularization algorithm.

The curves presented in Fig. 1 demonstrate the effectiveness of using regularisation based on the analysis of changes in the entropy of the probability distribution. This method improves the robustness of the Bayesian algorithm, which is used to determine the probability that a student belongs to a certain pattern (set of formed competences). Entropy analysis allows to identify moments when a student gives an answer that does not correspond to the expected level of his/her competences. In such situations, the system repeats the task with the same complexity characteristics as the previous one. This is similar to asking a clarifying question during a face-to-face

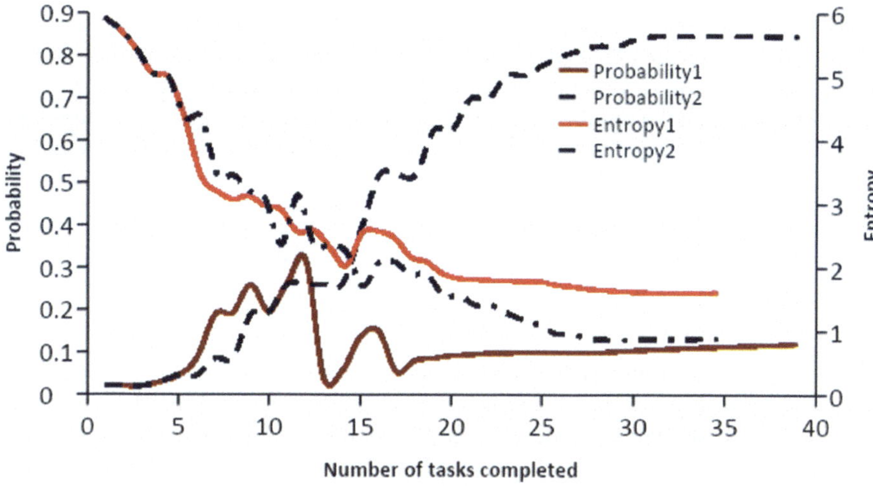

Fig. 1 Dependence of the probability of a student belonging to a given pattern and the entropy of the probability distribution over patterns with and without regularization (Probability2, Entropy2 curves) and without it (Probability1, Entropy1 curves) on the number of tasks completed

examination. Application of this approach makes it possible to assess the level of three competences on a four-point scale in one testing session by performing no more than two dozen tasks. The proposed algorithm imitates the process of competence assessment by an experienced examiner. If a student unexpectedly fails a simple task or unexpectedly succeeds in a very complex task, the system asks him/her an additional question from the same field to make sure that the results obtained are reliable. This approach increases the accuracy of competence assessment, making it more objective and reliable.

The above features of the implementation of the Bayesian algorithm made it possible to make this algorithm more stable and its results more adequate. To assess the reliability and accuracy of the results obtained by various methods, numerical experiments were carried out on a simulation model. Each experiment consisted of a multiple run—imitation of the testing process with the same parameters of the tests under study. In our experiments, 100 runs were performed. As a measure of the accuracy of estimating the studied characteristics, the standard deviation of this characteristic σ, calculated for all runs, is used.

100 runs of the Bayesian algorithm for evaluating a student with a well-developed first competence and a satisfactory level of other competences ($\theta 1 = 3$, $\theta 2 = 1$, $\theta 3 = 1$) showed that the average value of the probability of correct assessment (the algorithm assigns the student to the pattern to which this student belongs) increases monotonically with each completed task. However, even after completing 60 tasks, the probability of correct evaluation does not exceed 50%. In addition, the scatter of the probability value by runs is commensurate with the average value, which shows the low reliability of the method.

Taking into account the change in entropy increases the reliability of the method. 100 runs of the Bayesian algorithm show that a 70% probability of a correct assessment is achieved after completing 20 tasks. The spread in runs does not exceed 10% and depends little on the number of completed tasks. All this testifies to the sufficient stability of the regularized Bayesian algorithm and the possibility of its application to assess several competencies at once.

4 Conclusion

This paper presents an approach that allows using multivariate Rasch model to assess the level of formation of several competences in one testing session. This approach was tested on a simulation model of computer-based testing, where classical and adaptive testing methods using a Bayesian algorithm were compared. The instability of the algorithm was found to be due to the random nature of the performance of the task selected by the maximum of the information function. To address this problem, it is proposed to account for atypical changes in the entropy of the probability distribution of affiliation probabilities. Statistical analysis of the experimental results has shown that the regularised algorithm provides high accuracy of the assessment of the level of competence formation after performing several tasks. The proposed model makes it possible to measure the level of several competences in one testing session and is well adapted to the use of adaptive testing technologies. The use of intelligent systems of task selection in combination with the Bayesian algorithm allows to significantly reduce the number of tasks in the test while maintaining a high level of accuracy of competence assessment. Besides, the proposed model allows simulating the real process of a student's face-to-face assessment by a teacher, identifying 'illogical' answers and asking clarifying questions when necessary.

The development of computer testing systems using the proposed approach will allow teachers to more rationally organize the process of testing knowledge, get more information about the level of formation of students' competencies and thereby compensate for the shortcomings of distance learning, which we are forced to conduct in a pandemic.

References

1. UNESCO: Review of high-stakes exams and assessments during COVID-19. Available via DIALOG. https://en.unesco.org/sites/default/files/unesco_review_of_high-stakes_exams_and_assessments_during_covid-19_en.pdf. Accessed 25 Feb 2022
2. Gusyatnikov VN, Bezrukov AI, Sokolova TN, Kayukova IV (2015) Information technology to assess the level of competence in the educational process. In: Paper presented at the 9th international conference on application of information and communication technologies, AICT, pp 473–476

3. Yeung CK (2019) Deep-irt: Make deep learning-based knowledge tracing explainable using item response theory. arXiv preprint arXiv:1904.11738
4. Haberman SJ (1977) Maximum likelihood estimates in exponential response models. The Annals of Statist 5(5):815–841
5. Dempster AP, Laird NM, Rubin DB (1977) Maximum likelihood from incomplete data via the em algorithm. J of the Royal Statist Society: Ser B (Methodolog) 39(1):1–22
6. Geng J, Shi W, Hu G (2021) Bayesian nonparametric nonhomogeneous Poisson process with applications to USGS earthquake data. Spatial Statist 41
7. Wilson K, Karklin Y, Han B, Ekanadham C (2016) Back to the basics: Bayesian extensions of IRT outperform neural networks for proficiency estimation. In: Paper presented at the proceedings of the 9th international conference on educational data mining, pp 539–544
8. Partchev I (2004) A visual guide to item response theory. Friedrich-Schiller-Universitat, Jena. Available via DIALOG. https://docplayer.net/20748000-A-visual-guide-to-item-response-theory.html. Accessed 22 Dec 2021
9. Liu Y, Hu G, Cao L, Wang X, Chen M-H (2019) A comparison of monte carlo methods for computing marginal likelihoods of item response theory models. J of the Korean Statisti Society 48(4):522–523
10. Kibzun AI, Inozemtsev AO (2014) Using the maximum likelihood method to estimate test complexity levels. Autom Remote Control 75:607–621. https://doi.org/10.1134/S000511791404002X
11. Smirnov E, Dvoryatkina S, Martyushev N, Shcherbatykh S (2023) Software Package to Support Students' research activities in the hybrid intellectual environment of mathematics teaching. Mathematics 11(4):952
12. Pan J-C, Huang G-H (2014) Bayesian inferences of latent class models with an unknown number of classes. Psychometrika 79(4):621–646
13. Martyushev N, Shutaleva A, Malushko E, Nikonova Z, Savchenko I (2021) Online communication tools in teaching foreign languages for education sustainability. Sustainability (Switzerland) 13(19):11127
14. Natesan P, Nandakumar R, Minka T, Rubright JD (2016) Bayesian prior choice in IRT estimation using MCMC and variational Bayes. Frontiers in psychology
15. Shutaleva A, Starostin A, Nikonova Z, Savchenko I (2022) Migration potential of students and development of human capital. Education Sciences 12(5):324

Correction to: Structural and Technological Transformation of Education in the Post-Pandemic Period

Alexei L. Semenov, Vadim V. Grinshkun, Svetlana N. Dvoryatkina, and Vladimir A. Faerman

Correction to:
A. L. Semenov et al. (eds.), *Structural and Technological Transformation of Education in the Post-Pandemic Period*, Lecture Notes in Networks and Systems 1282, https://doi.org/10.1007/978-3-031-84039-5_18

In the original version of the book, the editor's name was inadvertently published incorrectly which has now been changed from Vadim V. Grishkun to Vadim V. Grinshkun. The book has been updated with the change.

The updated version of this book can be found at
https://doi.org/10.1007/978-3-031-84039-5

A. L. Semenov et al. (eds.), *Structural and Technological Transformation of Education in the Post-Pandemic Period*, Lecture Notes in Networks and Systems 1282,
https://doi.org/10.1007/978-3-031-84039-5_18

The manufacturer's authorised representative in the EU is Springer Nature Customer Service Centre GmbH, Europaplatz 3, 69115 Heidelberg, Germany. If you have any concerns regarding our products, please contact ProductSafety@springernature.com

Printed and bound by CPI Group (UK) Ltd, Croydon, CR0 4YY
01/10/2026
02225961-0001